CAMBRIDGE COMPARATIVE PHYSIOLOGY

GENERAL EDITORS:

J. BARCROFT, C.B.E., M.A.
Fellow of King's College and Professor of
Physiology in the University of Cambridge

and

J. T. SAUNDERS, M.A.
Fellow of Christ's College and Demonstrator
of Animal Morphology in the University of
Cambridge

COMPARATIVE PHYSIOLOGY OF THE HEART

COMPARATIVE PHYSIOLOGY

OF

THE HEART

BY

A. J. CLARK, M.C., M.D.

Professor of Materia Medica in the University
of Edinburgh, formerly in the University
of London, University College and
in the University of
Cape Town

CAMBRIDGE

AT THE UNIVERSITY PRESS

1927

CAMBRIDGE
UNIVERSITY PRESS

University Printing House, Cambridge CB2 8BS, United Kingdom

Cambridge University Press is part of the University of Cambridge.

It furthers the University's mission by disseminating knowledge in the pursuit of education, learning and research at the highest international levels of excellence.

www.cambridge.org
Information on this title: www.cambridge.org/9781107502444

First published 1927
First paperback edition 2015

A catalogue record for this publication is available from the British Library

ISBN 978-1-107-50244-4 Paperback

PREFACE

The author has endeavoured to discuss within the limits of a small volume certain aspects of the heart's functions, which appear to be of particular interest when viewed from the standpoint of comparative physiology.

Tigerstedt's *Physiologie des Kreislaufes* (4 vols. 1921–23) gives an exhaustive account of the comparative physiology of vertebrate circulation and also deals at considerable length with the invertebrate circulation. Von Brücke (Winterstein's *Handbuch der Vergleichenden Physiologie*, vol. I. i. 1925, pp. 827–1108) has given a full account of the comparative physiology of the circulation in Invertebrates and in cold-blooded Vertebrates. These excellent and recent works have made it unnecessary to devote much space to the comparative morphology of hearts, and the author has avoided reproducing illustrations which appear in these works. They also provide exhaustive bibliographies and therefore only selected references have been given in this volume.

The greater portion of the volume has been devoted to the consideration of the comparative physiology of the vertebrate heart. This is due to the fact that information concerning the physiology of invertebrate hearts is so scanty that it is difficult to make any generalisations from the data available.

Even in the case of the hearts of warm-blooded Vertebrates, accurate information is only available concerning a few hearts and the writer is aware that many of the generalisations attempted are somewhat hazardous. He can only hope that the speculative nature of some of the deductions in this volume may stimulate research workers to provide more extensive and accurate information in this field of comparative physiology.

The writer desires to express his thanks both to Dr J. Beattie for providing much valuable data regarding vertebrate hearts from the post-mortem room of the menagerie of the Zoological Society and to this Society for the facilities which they have provided.

A. J. C.

1927

CONTENTS

Chapter I

ECHINODERMATA AND VERMES

Echinodermata—Vermes—Hirudinea—Annelida—
Peristalsis of Blood Vessels.

A mechanism for the circulation of body fluids is necessary, even in relatively simple forms of life, since any animal which has specialised organs for respiration, digestion or excretion requires some form of circulation. The circulation performs a very large number of functions, but the supply of oxygen to the tissues, and in particular its supply to the contractile tissues, is the most urgent and important function of the circulation in most animals. In general, the power of an animal to perform continuous muscular movements depends very largely upon the efficiency of the circulation, and active, quick-moving animals require a more complex and efficient circulatory system than animals of sluggish and sedentary habits. The activity of even closely related species of animals varies very greatly, and hence great variations in the complexity and efficiency of the circulatory mechanisms are found even in animals that are nearly related. This variation is particularly well marked in the Invertebrates, but the limits of space only allow of the description of a few examples of the better developed types of hearts occurring in each class of Invertebrates, and it must be remembered that in each class there are plenty of examples of much less perfect circulatory systems.

ECHINODERMATA

Rhythmic contractility is a property which occurs very frequently in undifferentiated protoplasm, and all movements tend to shift the body fluids. The simplest system that is definitely circulatory is seen in the Echinodermata. The Holothurians have vessels in connection with the gut which alternately contract and relax; these contractions are not propagated, and serve simply to mix up the blood in the vessels. The process is very like the pendulum movements seen in the vertebrate intestine.

c

VERMES

The Worms are the lowest class of animals in which any structure analogous to a heart is found. The development of the circulation varies enormously within the class. The Platyhelminthes and Nemathelminthes either have no true circulatory system or only a very rudimentary one. The Hirudinea and Annelida have well-developed circulatory systems, and a certain amount of information is available concerning their physiology.

Circulation in Hirudinea

The circulation in Hirudinea is of the open type; there is a system of vessels through which the blood circulates, but this system communicates with large sinuses. Gaskell(2) has described the circulation in *Hirudo*. He found that peristaltic waves passed down the lateral blood vessels about 6 times a minute. The two lateral vessels contracted alternately and the blood was thus driven from side to side of the animal.

Bidder noted in *Nephelis* that the peristaltic waves usually passed down the lateral vessels from behind forwards, but that occasionally the peristalsis was reversed.

Circulation in Annelida

Many of the Annelids possess a well-developed circulatory apparatus. The circulation is of the closed type, and the blood is driven out from a contractile vessel through arteries to a closed capillary system, from which it is returned to the contractile vessel by veins.

The circulation in *Lumbricus* has been studied extensively. The dorsal vessel alone is contractile, and in this regular peristaltic waves pass from behind forwards and drive the blood forwards and also outwards through lateral vessels. The blood passes through capillaries and is collected in the ventral vessel from which it returns to the dorsal vessel. The exact course of the circulation is a matter of dispute.

In many of the Polychaeta a further development of the circulatory system is found. The general arrangement of the circulation is similar to that of *Lumbricus*, but in addition the anterior portion of the dorsal vessel is expanded to form a cavity with relatively thick walls. This may be regarded as the most primitive form known of a true heart.

PERISTALSIS OF BLOOD VESSELS

The dorsal contractile vessel in *Lumbricus* is very easy to observe through the skin of the worm. At room temperature (15° C.) I noted that contractions occurred 15–20 times a minute and were propagated forwards at a rate of about 20 mm. per second. The duration of the contraction at one spot was about 1 sec. and the length of the wave of contraction was about 20 mm.

Stübel (3) has made a careful examination of the peristaltic activity of the dorsal vessel of *Lumbricus*. In the normal worm the peristaltic waves always start from the anal end and run forwards to the end of the dorsal vessel. If the worm is cut in two, regular peristalsis continues in both halves and in both halves the waves run from behind forwards. When the worm is divided into a number of segments the peristalsis continues in each segment.

The peristaltic activity is, however, largely dependent on the filling of the vessel with blood, and ceases if the vessel is emptied. The frequency of peristalsis is increased by raising the temperature. The peristaltic activity appears to be completely independent of nervous control, for it is not affected by removal of the ventral nerve strand. Nerve cells are, however, scattered around the dorsal vessel of *Lumbricus* and therefore it is impossible to exclude nervous control with certainty.

The dorsal vessel of *Nereis* shows the same type of activity as that of *Lumbricus*. I found at room temperature that the rate was 8 per minute and that the rate of conduction was 10–20 mm. per second.

Carlson (1) examined the circulatory system in *Arenicola* and *Nereis*. He found that isolated portions of the dorsal vessel and the isolated oesophageal heart showed rhythmic contractions and that they responded by contraction to electrical stimuli. He found that during systole these tissues showed a relative refractory period in which their sensitivity to electrical stimuli was diminished. Continuous stimulation of the vessels produced, however, a local tetanus or tonus, and therefore the refractory period was only poorly marked. He found that artificially induced contractions could be conducted in either direction.

Carlson noted ganglion cells scattered around the dorsal vessels and heart, and observed that in *Arenicola* stimulation of the ventral nerve cord produced inhibition of the oesophageal heart, but increased the frequency of the dorsal vessel. He admitted, however, that these effects might be in part secondary to the contractions of the gut muscle produced by stimulation of the ventral cord.

The dorsal vessels and hearts of the Worms therefore show some of the properties which are regarded as the fundamental properties of heart muscle. All portions of the system show spontaneous rhythmic activity, but this is greatest at the posterior end and the excitation is conducted down the length of the heart. Contraction appears to be followed by a relative refractory period during which time irritability is somewhat reduced.

In the Polychaeta the division into dorsal vessel and oesophageal heart is analogous to the division in higher animals into auricle and ventricle. The fundamental properties of spontaneous rhythmicity, conductivity and reduction in excitability immediately after activity appear to be very widespread properties in all forms of protoplasm capable of contractile activity. There does not appear to be any evidence to show that these properties are of necessity dependent on nervous connections.

The properties of the dorsal vessels of the Worms are of particular interest because they show a considerable resemblance to the properties of the embryonic hearts of Arthropods and Vertebrates, in both of which the heart develops as a simple tube of muscle cells.

References

(1) Carlson. *Amer. Journ. of Physiol.* 22, 353. 1908.
(2) J. F. Gaskell. *Proc. Roy. Soc.* B, 205, 153. 1914.
(3) Stübel, H. *Pflüger's Arch.* 129, 1. 1909.

Chapter II

HEARTS OF ARTHROPODA

Xyphosura—Crustacea—Arachnida—Myriapoda—Insecta—
Properties of Arthropod Heart Muscle.

The Arthropods all have a similar and peculiar type of circulation characterised by an ostiate heart. The heart is surrounded by a blood sinus in which the venous blood collects. The blood passes from the venous sinus into the heart by means of ostia in the walls of the heart. When the heart contracts, the ostia are closed and the blood is driven out of the heart through vessels. Arthropod hearts in most cases are segmented and the chambers are very frequently divided by means of valves.

The hearts of Arthropods are on the whole remarkably inefficient as compared with the high development of the other organs in these animals.

The development of the circulatory apparatus varies very greatly in the Arthropods, for naturally the gill-breathing Crustacea have a better developed circulation than the Insecta and Myriapoda, since the distribution of oxygen throughout the body by the tracheae in the latter forms makes their oxygen supply independent of the circulation, and in consequence the circulatory apparatus is very rudimentary even in tracheate animals which are capable of violent and sustained activity.

The Insecta therefore are a complete exception to the general rule that the development of the circulation is proportional to the capacity of the animal for sustained exertion.

The Arachnida are intermediate between the Crustacea and Insecta. These animals breathe partly by tracheae and partly by localised lungs, and their circulatory apparatus is better developed than that of the Insecta.

XYPHOSURA

The king crab, *Limulus*, has the most highly developed circulation of any Arthropod. In most Arthropods the heart drives blood to the organs and these organs are surrounded by venous sinuses which communicate with the venous sinus around the heart. In *Limulus* there is a closed system of veins emptying into a venous sinus around the heart.

The heart of *Limulus polyphemus* consists of a segmented tube 15–20 cm. long and 2·5 cm. broad. It is composed of 8 segments and in each segment there is a pair of ostia. The heart has 4 pairs of lateral arteries and at the anterior end there are cranial arteries.

On the dorsal surface there is a nerve chain which may be considered an elongated ganglion composed of ganglion cells and nerves. This chain is separated from the heart by the ectocardium and can be removed without injury to the heart.

Nukada (11) states that in *Limulus longispina* there is a thin-walled auricle which covers the posterior end of the heart and that the structure usually described as the whole heart of *Limulus* is really only the ventricle. He found that the heart beat originated in the auricle.

Carlson (2) has shown that the heart beat of *Limulus* is entirely dependent on and controlled by the dorsal ganglion.

When the ganglion is present the heart contracts rhythmically 12–16 times per minute. The contraction begins at the posterior end but the whole heart contracts almost simultaneously, the interval between the contraction of the two ends of the heart being about 0·025 sec. Removal of the ganglion arrests all automatic activity of the heart and so also does section of the nerves between the ganglion and the heart. When the heart muscle is divided transversely the two halves continue a co-ordinated beat, but if the ganglion is divided the two halves of the heart beat independently.

The heart beat in *Limulus* therefore is originated in the ganglion and is conducted by the nerves.

The heart when deprived of its dorsal ganglion behaves similar to a nerve-muscle preparation of the frog. Stimulation either of the nerves or the muscle causes a contraction, and this contraction is proportional to the strength of stimulus. The contraction is followed by a slight diminution in excitability but summation and tetanus can easily be produced.

A heart in good condition with the nerve ganglion intact shows more resemblance to the vertebrate heart, for its response to stimuli of varying intensities is more uniform than that of the heart deprived of the ganglion. Carlson's conclusions regarding the heart of *Limulus* appeared quite conclusive, but they have been challenged recently by Hoshino (8), who states that if the heart of *Limulus* be exposed without any injury to the suspensory ligaments, then section of the nerves does not arrest the heart, and concludes from this that myogenic rhythm and conduction are possible in the heart of *Limulus*. The evidence regarding the mode of transmission of the wave of excitation in the heart of *Limulus* is therefore contradictory, and the mode of the origination and conduction of the wave of excitation in this heart must be considered as still uncertain.

Nukada (11) and Hoffmann (7) have both published records of the electrical response of the *Limulus* heart. Hoffmann considered that the response was of an oscillatory nature, but Nukada denied this. Hoffmann found that a stimulus applied direct to the heart deprived of its ganglion produced an electrical response which consisted of a single oscillation, but that stimulation of the ganglion caused a series of oscillations.

Carlson (5) showed that the *Limulus* heart was supplied by augmentor and inhibitor nerves which passed from the brain to the dorsal ganglion. Stimulation of the brain or of the 7th or 8th haemal cranial nerves inhibited the heart, causing a diminution in amplitude and frequency of contractions. Augmentor nerves came from the abdominal ganglion and stimulation of the commissure between the brain and the first abdominal ganglion produced an increase in the frequency and amplitude of contraction. Carlson found that these nerves acted on the dorsal ganglion of the heart and not directly on the cardiac muscle.

The dorsal ganglion can also act as a reflex centre since increased filling of the isolated heart causes augmentation of the beat, an effect which is abolished if the connections with the dorsal ganglion are cut.

The development of the *Limulus* heart is of interest, for Carlson and Meek (6) showed that the heart developed in the embryo as a tube composed of a syncytium of unstriated muscle fibres. Automatic rhythmicity commenced at the 22nd–23rd day but nerve strands only appeared on the 28th–29th day. The *Limulus* heart therefore appears to develop as a tube of unstriated muscle capable of originating and conducting contractions and this is replaced during development by striated muscle which is completely under the control of the nervous system.

CRUSTACEA

The hearts of Crustacea consist of a sac or tube with venous ostia at the sides and arteries passing from the anterior and posterior ends. In the Isopods the heart is tubular and has several pairs of ostia, but in the Decapods it consists of a single chamber, with 4 ostia through which the blood enters, and with anterior and posterior arteries through which the blood is expelled. This sac is suspended in a venous sinus and the wall of this venous sinus is stated by Mangold (10) to be rhythmically contractile and thus to play the part of an auricle. The heart is composed of striated muscle and the rate of conduction is so rapid that all parts of the sac contract almost simultaneously. The contractions appear to originate in the portion around the venous ostia. The system is primitive and inefficient and the systolic arterial pressure in a lobster is only about 12 cm. of water (1).

In some Crustacea, e.g. some Isopods, the heart is tubular, extending the whole length of the body. In some classes, e.g. Ostracods and Copepods, most of the species have no heart and the blood is moved solely by the movements of the body. The larger Decapods have well-developed hearts and those of the lobster, cray-fish and crab have been investigated by numerous observers. Nothing resembling the dorsal ganglion of *Limulus* is seen in these hearts, although scattered ganglion cells may occur in the heart. The isolated hearts beat regularly when infused with a suitable fluid, and if a heart is cut up each portion continues to contract (Brandt, 1865).

There is no evidence therefore that the cardiac rhythm in the crustacean heart is of neurogenic origin. The heart is composed of striated muscle fibres, and the response appears to be of an oscillatory nature, quite unlike the electric response of the vertebrate heart. An example of the electrical response of the heart of *Homarus* is shown in Fig. 1. The impulse probably originates around the ostia, but on inspection the whole heart appears to contract simultaneously. I took electrical records from the centre and periphery of the heart and concluded that the rate of propagation of the impulse over the heart must be at least 400 mm. per second. The brothers Weber (1846) noted that the crustacean heart showed no refractory period and that summation and tetanus could easily be produced by electrical stimuli. This has been confirmed by numerous other workers.

Carlson (4) found that the hearts of *Cancer* and *Palinurus* gave an all or none response to electrical stimuli over a considerable range of intensity. The activity of the heart is largely conditioned by the internal pressure, and Carlson found that the heart of *Cancer in situ* beat 90–120 times a minute, but that after a considerable loss of blood the frequency sank to 15–30 per minute.

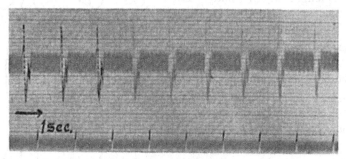

Fig. 1. Electrical response of lobster heart.

Similarly, the isolated heart beats irregularly when empty but recovers if internal pressure is applied.

Numerous observers have examined the nervous supply of crustacean hearts and there is general agreement that both inhibitor and augmentor nerves exist, and the centre of these nerves appears to be in the anterior portion of the thoracic ganglion.

ARACHNIDA

The spider's heart consists of a single chamber with 3 or 4 pairs of ostia. It is furnished with lateral, cranial and caudal arteries. The whole heart appears to contract simultaneously. Willem and Bastert (12) recorded the pulsations of the heart of a spider (*Pholeus*) and found the rate to be 100 per minute. The scorpions have a better developed heart which consists of numerous chambers each provided with ostia. Newport (1848) observed that each chamber of the heart contracted in turn from behind forwards, but Blanchard (1852) thought that the whole heart contracted simultaneously. Carlson (5) considers it probable that the hearts of spiders are provided with accelerator fibres, and also considers that there is some evidence for the existence of inhibitor nerves.

MYRIAPODA

The hearts of Myriapoda resemble those of Insects and consist of segmented tubes lying along the dorsum of the animal. The segments may number 160 (Newport). The heart runs the whole length of the body. Each segment is provided with venous ostia and lateral arteries. Newport described the heart as consisting of an outer tunic and an inner true heart. The arrangement is shown in Fig. 2.

INSECTA

The hearts of Insects resemble those of the Myriapoda. The mode of function appears to vary however. In the larvae the contraction is a peristaltic wave passing slowly down the heart from behind forwards (Milne-Edwards, Dogiel), and when the heart is divided each portion continues to beat (Weber). Lasch (9) observed a rate of conduction of 19–44 mm. per second in the *Lucanus* larva. Dr Seliškar kindly made some observations for me on various insect hearts. The passage of the wave of contraction in the larva of *Cossus* is shown in Fig. 3. The heart of the adult *Carausius* (stick insect) was found to function in an exactly similar manner, and a wave of contraction travelled along the heart at about 20 mm. per second. In *Periplaneta Americana*, however, I found that all the chambers of the heart appeared to

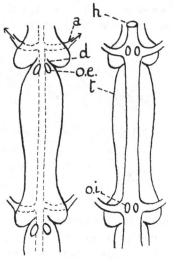

Fig. 2. Heart chamber of *Scolopendra*. *a.* artery; *d.* valve; *h.* inner true heart; *t.* external tunic; *o.e.* ostia in external tunic; *o.i.* ostia in inner true heart. (From figure by Newport.)

contract simultaneously and Carlson observed the same in a grass-hopper's heart which was 4 cm. long.

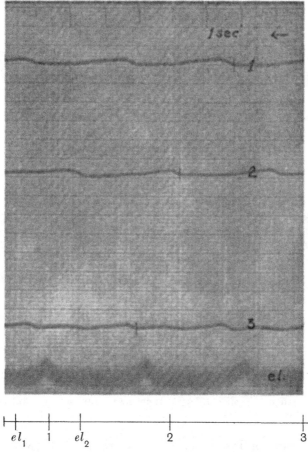

Fig. 3. Mechanical and electrical response of heart of larva of *Cossus cossus*. The positions of the leads are shown below the figure. Lead 1 is at the anal end. el_1, el_2, electrode leads; 1, 2, 3, mechanical leads. The distances between the leads were: el_1–el_2, 5 mm.; el_1–1, 10 mm.; 1–2, 32 mm.; 2–3, 27 mm. The rate of transmission of the wave of contraction was 24 mm. per second. (Dr A. Seliškar.)

The records of electrical responses given in Figs. 3 and 4 show that the response of the larval heart of Insects is completely unlike that of the crustacean heart. In the larval heart there is no suggestion of oscillations, and the records obtained show a simple diphasic response very similar in character to those obtained from vertebrate smooth muscle.

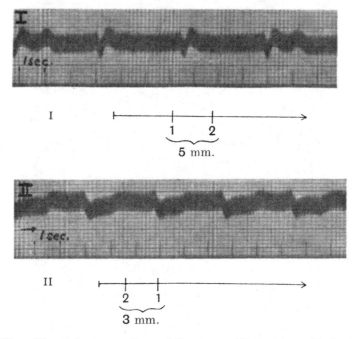

Fig. 4. Electrical response of larva of *Cossus cossus*. The positions of the leads are shown below the figures. The arrows show the direction of the wave of contraction. Negative deflection is downwards in I and upwards in II. (Dr A. Seliškar.)

The hearts of Insects appear therefore to be of two distinct types, and it is possible that as a general rule in Arthropods the heart is laid down in the embryo as a tube of smooth muscle and that this is replaced by striped muscle. This is known to occur in *Limulus*, and in Crustacea the response of the heart resembles that of striped muscle. In Insects it appears that the larval hearts resemble smooth muscle in their response, whereas the hearts of adult Insects vary. In most Insects there is a rapidly conducted contraction resembling that of striped muscle, whereas the heart of the adult stick insect resembles the larval heart.

Carlson (3) found that the heart of *Dictyphorus* (grasshopper) continued to beat when excised from the body, and that when divided the pieces continued to contract. Brandt (1866) observed the same fact in a number of different Insects.

The influence of the nervous system on the insect heart appears doubtful. Lasch observed that in some cases stimulation of the crop ganglion in the *Lucanus* larva caused inhibition of the heart, and Carlson found that stimulation of the head ganglion of *Dictyphorus* caused acceleration of the heart. Ganglion cells occur scattered round the heart, but there is no evidence that they play any part in the circulation or conduction of excitation.

THE DISTINCTIVE PROPERTIES OF ARTHROPOD HEART MUSCLE

Arthropod hearts differ very markedly from the hearts of all other animals, for in the first place they are composed of striated muscle fibres which do not form a syncytium. Moreover, the arrangement of the ventricle lying in a venous sinus and being filled through ostia is peculiar to Arthropods.

Carlson and Meek's observation that the heart of the *Limulus* embryo is composed of unstriped muscle fibres forming a syncytium, suggests that the heart of the adult Arthropod is a specialised structure differentiated from a form resembling the tubular hearts of Worms. This view is supported by the fact that although the heart of insect larvae is composed of striped muscle, yet functionally it resembles closely the worm's heart, for its contraction consists of a peristaltic wave passing slowly down the tube.

Information regarding the hearts of adult Arthropods is scanty, but it appears that every type of heart exists intermediate between the caterpillar heart and the *Limulus* heart.

In some adult Insects the contraction consists of a slow peristaltic wave, whilst in others the whole heart contracts simultaneously.

In the Arachnida and Myriapoda the contraction appears to consist of a peristaltic wave, whilst in the Crustacea the whole heart appears to contract simultaneously. Finally, in *Limulus* the whole heart contracts almost simultaneously and the origination and conduction of the beat are neurogenic.

Our present knowledge is insufficient to permit a decision as to the myogenic or neurogenic nature of the beat in Arthropods other than *Limulus*. In the latter case the evidence for neurogenic origin and conduction of contraction is very strong, but the circulation in *Limulus* appears to be altogether exceptional, and nothing resembling the *Limulus* heart ganglion has been found in any other Arthropod. It is of course tempting to assume that the slow peristaltic type of contraction, which so closely resembles the contraction of plain muscle, is myogenic in nature, and that the rapidly conducted type of contraction is neurogenic in nature. The evidence at present available is, however, insufficient to justify any such generalisation. In view of the large amount of attention that has been paid to the *Limulus* heart it is, however, necessary to emphasise the fact that the properties of this heart are peculiar and in no way typical of Arthropod hearts as a whole.

References

(1) Brücke and Satake. *Zeit. f. all. Physiol.* **14**, 28. 1912.

(2) Carlson. (*Limulus* Heart.) *Amer. Journ. of Physiol.* **12**, 67, 471. 1904.

(3) Carlson. (Insect Heart.) *Amer. Journ. of Physiol.* **14**, 6. 1905.

(4) Carlson. (Crustacean Heart.) *Amer. Journ. of Physiol.* **15**, 127. 1905.

(5) Carlson. (Properties of Invertebrate Heart Muscle.) *Ergebn. d. Physiol.* **8**, 371. 1909.

(6) Carlson and Meek. *Amer. Journ. of Physiol.* **21**, 1. 1908.

(7) Hoffmann. *Arch. f. (Anat.) Physiol.* p. 135. 1911.

(8) Hoshino. *Pflüger's Arch.* **208**, 245. 1925.

(9) Lasch. *Zeit. f. all. Physiol.* **14**, 312. 1913.

(10) Mangold. *Zeit. f. Wiss. Biol.* C, **2**, 184. 1924.

(11) Nukada. *Mitt. a. d. Med. Fak. d. Kaiserl. Univ. z. Tokio*, **19**, 1. 1917.

(12) Willem and Bastert. *Arch. Néerl. de Physiol.* **2**, 287. 1917.

Chapter III

MOLLUSCA AND TUNICATA

Mollusca—Properties of Heart Muscle—Nervous Control of Heart—
Circulation—Tunicata.

The Molluscs show every variety of heart development. In some
Scaphopods there is only a rudimentary heart connected by ostia
with the blood sinuses, whereas the hearts of the Cephalopods
are as complex and efficient as the hearts of Fishes or Amphibia.

The typical arrangement consists of a heart composed of a
ventricle and a varying number of auricles. The auricles contract
first and fill the ventricle, and from the ventricle the blood is
driven through arteries and capillaries into veins; these veins
usually break up into a second series of capillaries in the gills or
lungs, and from the lungs the blood is returned to the auricles.

The arrangement of a few typical hearts is shown diagrammatically in Fig. 5, which has been adapted from a diagram by

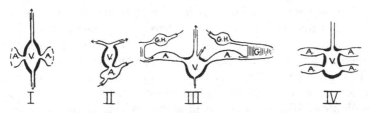

Fig. 5. Molluscan hearts. *V.* ventricle; *A.* auricle; *G.* gills; *G.H.* gill heart.
I. Lamellibranchiata. II. Gastropoda. III. Cephalopoda (dibranchiata).
IV. Cephalopoda (tetrabranchiata).

Griffiths(9). The circulation in Molluscs is on the whole much
more highly developed than in Arthropods. The molluscan ventricle
is usually a powerful muscle capable of producing a considerable
pressure. This is to be expected since the ventricle has to drive
the blood through a double system of capillaries.

The circulation is most highly developed in the Cephalopods,
which have accessory gill hearts to drive the blood through the
gills. The *Octopus* heart can produce a pressure of from 25–80 mm.

Hg, which is a higher blood pressure than that occurring in many cold-blooded Vertebrates.

PROPERTIES OF MOLLUSCAN HEART MUSCLE

In all Molluscs the heart is composed of plain muscle and the fundamental properties appear similar in all classes. On the whole they resemble the properties of vertebrate plain muscle, but in some respects approach those of vertebrate cardiac muscle. In molluscan hearts in general Carlson[4] found evidence of the existence of a relative refractory period at the beginning of systole, during which time the minimal intensity of stimulus needed to produce a response was greater than normal. A sufficiently strong stimulus would, however, produce a response at any point in the cardiac cycle, and hence tetanus could be produced in all molluscan hearts provided a sufficiently strong stimulus was applied. The relative refractory period was better marked in the higher than in the lower species of Mollusca.

Carlson[3] also found that the response of molluscan hearts was not proportional to the strength of stimulus over a wide range and that hearts in good condition responded with a contraction of uniform strength to stimuli of increasing intensity over a wide range, but that an increase of the intensity of stimulus above this range was followed by contraction of increased strength accompanied by a more or less prolonged tonic contraction.

The properties of the molluscan heart muscle therefore resemble those of vertebrate plain muscle rather than those of vertebrate cardiac muscle, but in some respects show characters intermediate between the two types of muscle.

Records of the electrical response of the molluscan heart are unfortunately very scanty, but those obtained by Evans[6] from the isolated heart of *Helix* show that the electrical response is not oscillatory but is a slow prolonged diphasic variation.

Inspection of the heart of *Pecten in situ* shows that there is a well-marked auriculo-ventricular interval of about 0·5 sec. The electrical records of the heart of *Helix* (Evans) and of the oyster (Eiger[5]) show similar a.-v. intervals. The frequency of molluscan hearts is low; the following are typical frequencies observed in hearts at room temperature.

Table 1

LAMELLIBRANCHS

Pecten ...	...	22	(Clark)
Mytilus ...	...	10–15	(Carlson)
Mya arenaria	...	14	(Yung)
Cardium	...	15–17	(Carlson)
Anodon ...	...	2–4	(Koch)

GASTROPODS

Helix pomatia	...	20–40	(Yung)
Pterotrachea	...	50–80	(Knoll)
Ariolimax	...	35–40	(Carlson)
Aplysia ...	...	33–34	(Botazzi and Enriques)
Pleurobranchia ...		20–30	(Carlson)

CEPHALOPODS

Loligo ...	...	70–80	(Carlson)
Octopus ...	...	35–40	(Botazzi and Enriques)
Sepia ...	...	40	(Bert)

All parts of the heart appear to show automatic contractions. Rywosch[13] noted that in *Pterotrachea* the ventricles contracted when the auricles were arrested, and Ransom[12] found that in the *Octopus* even small pieces of ventricle contracted spontaneously. The frequency of contraction appears, however, to depend very largely upon internal pressure. The following figures were found by Biedermann for the heart of *Helix*:

Internal pressure in mm. water ...	30	15	8	5	2	30
Frequency per minute	50	36	21	11	0	50

This relation between internal pressure and frequency has been noted by most observers who have worked with the isolated hearts of Molluscs. Increase of pressure causes an increase of frequency in most hollow organs whose musculature shows a spontaneous rhythm. The relation is particularly well marked in the molluscan heart but a similar effect can be demonstrated in the hearts of Worms, Arthropods and Vertebrates.

THE NERVOUS CONTROL OF THE MOLLUSCAN HEART

This is difficult to demonstrate because the frequency and force of response of the molluscan heart depend on the filling of the heart. Stimulation of nerves usually produces muscular contraction and this may cause increased filling of the heart and thus appear to produce an augmentor effect.

Molluscan hearts closely resemble plain muscle in their general

c

properties and the evidence available suggests that the origin and conduction of the excitation are purely myogenic.

Dogiel and Carlson (1, 2) both describe nerve ganglion cells in a number of molluscan hearts. Many recent observers have, however, failed to find such cells, and this is evidence that they cannot be plentiful and that no extensive local nerve plexus can exist.

A. *Lamellibranchs*

Carlson studied the hearts of *Mytilus*, *Mya*, *Platydon*, *Venus*, *Cardium*, *Hennites* and *Pecten*, and in all of these found that stimulation of the nerves leading from the visceral ganglion to the heart caused a well-marked inhibition of the auricle and of the ventricle.

In *Mytilus* Carlson found that stimulation of the cardiac nerves had no certain action on the heart.

B. *The Chitons*

Carlson concluded that the chief action of the cardiac nerves in *Chiton* was augmentor and accelerator.

C. *Prosobranchiata*

Carlson found that in *Haliotis*, *Sycotypus*, *Natica* and *Lucapina*, the nerves had only an augmentor action, and that there was no evidence for the existence of depressor nerves.

D. *Tectibranchiata*

Dogiel (1876) described augmentor nerves in *Aplysia*, and Botazzi and Enriques (1901) concluded that augmentor nerves alone were present. Carlson examined *Aplysia*, *Bulla* and *Pleurobranchia* and found that in all cases only augmentor nerves were present.

E. *Nudibranchiata*

Carlson examined *Montereina* and *Triopha* and in' the latter case found that stimulation of the visceral nerves produced at first an augmentor action and later on a depressor action.

F. *Pulmonata*

Yung (1881) and Ransom (1884) both showed the existence of inhibitor nerves in *Helix*, and this observation was confirmed by Carlson, and extended to *Ariolimax* and *Limax*. He concluded that stimulation of the visceral nerves usually produced inhibition, but that less frequently augmentor effects were produced.

Carlson concluded that the Gastropods show an ascending complexity in the nervous control of the heart. In the lower Gastropods (Prosobranchs and Tectibranchs) the heart nerves have a purely augmentor action. In one Nudibranch, *Triopha*, there is clear evidence of a double innervation, augmentor and inhibitor, but the inhibitor function is less marked than the augmentor. In the higher Gastropods (*Helix* and *Limax*) there is a double effect, but the inhibitor effect is more marked than the augmentor.

Furthermore, in *Limax* the nerves have a double effect on the auricle but only a depressor effect on the ventricle, whereas in *Helix* the nerves have a double effect on both the auricle and ventricle.

G. *Cephalopods*

Various authors have shown the existence of inhibitory nerves to the hearts and gill hearts of *Octopus* and *Sepia*.

MOLLUSCAN CIRCULATION

Willem and Minne (15) measured the pressure in the ventricle, auricle and aorta of *Anodonta* and found that the pressure in the auricle was about 5 mm. water and that in the ventricle and aorta was 10 mm. water during diastole, and rose to 30 mm. water during systole. The pressures produced by the gastropod heart are considerably greater than the above.

Bethe found that the pressure inside the body cavity of a resting *Aplysia limarina* was 2·5–4·0 cm. water, and that on movement it rose to as much as 6·0 cm. The average pressure is probably about 3·0 cm. water. Straub (14) considers that the heart is protected from this pressure in part by the pericardium, and the ventricle must therefore drive the blood through the arteries against this pressure. Straub found that, when the isolated heart of *Aplysia* contracted isometrically, the pressure during contraction rose with the initial tension and that the optimal initial tension was between 20 and 40 mm. Hg. These figures show that the *Aplysia* ventricle is a powerful organ.

The cephalopod heart is much the most complex and powerful found in the Invertebrates. The circulatory apparatus consists of two auricles which receive the blood from the gills and a ventricle which drives the blood round the body capillaries. The venous blood passes to the gills and is driven through them by two branchial hearts.

Fuchs[8] found that the aortic pressure varied from 25 to 80 mm. Hg and that the difference between diastolic and systolic pressures was usually 10 mm. but might amount to 25 mm. Hg.

Fredericq[7] found that the pressure in the gill veins was 7–8 cm. blood (5–6 mm. Hg) and in the aorta varied from 62 to 78 cm. blood (48–60 mm. Hg), and in one case in a large *Octopus* amounted to 88 mm. Hg. These figures for the systolic pressure are very remarkable since they are greater than those found in most cold-blooded Vertebrates.

TUNICATA

The tunicate heart is a long tubular structure down which travel peristaltic waves. In *Phallusia* the heart is about 10 cm. long in a medium-sized specimen. The author observed a frequency (at 12° C.) of about 10 per minute and the waves travelled at a rate of between 10 and 20 mm. per second. Carlson[1] observed a rate of 20–35 mm. per second in the heart of *Ciona*.

These hearts show a considerable general resemblance to those of Worms in their activity. The striking peculiarity of tunicate hearts is that their rhythm is reversible. One end acts as a pacemaker for a number of beats and waves of contraction run in one direction, and then the other end acts as a pacemaker and the contraction travels in the reversed direction for a somewhat similar period. Usually the number of advisceral contractions is greater than the number of abvisceral. This reversal of rhythm occurs in the isolated heart (Hecht[10]).

Nicolai[11] found that if the heart was cut in two the two ends contracted, the contraction waves running towards the middle. He also found that quite small pieces of the heart could show automatic rhythmicity.

The author has observed in the rat's uterus, both isolated and *in situ*, a shifting of the point of origin of automatic contractions. In this case the site of origin shifted in an irregular fashion up and down the tube, but the observation suggests that, in any tube which is not so definitely polarised that one portion has a much greater automatic rhythmicity than the rest of the tube, the site of origin of the contractions is likely to shift. In the tunicate heart presumably the automatic rhythmicity of the two ends is nearly equal and considerably greater than that of the rest of the tube.

References

(1) Carlson. (General Properties of Molluscan Hearts.) *Ergebn. der Physiol.* **8**, 371. 1909.

(2) Carlson. (Cardiac Nerves in Molluscs.) *Amer. Journ. of Physiol.* **13**, 396, and **14**, 16. 1905.

(3) Carlson. ("All or none" Response.) *Amer. Journ. of Physiol.* **16**, 85. 1905.

(4) Carlson. (Refractory Period.) *Amer. Journ. of Physiol.* **16**, 67. 1905.

(5) Eiger. *Pflüger's Arch.* **151**, 1. 1913.

(6) Evans. *Zeit. f. Biol.* **59**, 29. 1912.

(7) Fredericq. *Arch. Intern. de Physiol.* **14**, 126. 1914.

(8) Fuchs. *Pflüger's Arch.* **60**, 173. 1895.

(9) Griffiths. *Physiology of the Invertebrates.* 1892.

(10) Hecht. *Amer. Journ. of Physiol.* **45**, 157. 1918.

(11) Nicolai. *Arch. f. Physiol.* Suppl. 87. 1908.

(12) Ransom. *Journ. of Physiol.* **5**, 261. 1884.

(13) Rywosch. *Pflüger's Arch.* **109**, 355. 1905.

(14) Straub. *Pflüger's Arch.* **103**, 429. 1904.

(15) Willem and Minne. *Mém. de l'Acad. Roy. de Belg.* **57**. 1898.

Chapter IV

GENERAL PHYSIOLOGY OF HEARTS OF COLD-BLOODED VERTEBRATES

Cyclostomata—Elasmobranchii—Teleostei— Dipnoi—Amphibia—
Reptilia—Coronary Circulation—Heart Valves—Lymphatic Hearts.

The hearts of all Vertebrates are of a similar type and are the result of the differentiation of a simple tubular heart into a series of chambers.

CYCLOSTOMATA

The heart of the Cyclostome (e.g. *Bdellostoma*) represents a fairly simple development of a tubular heart, and consists of a sinus venosus, auricle, ventricle and bulbus arteriosus.

The large veins open into the thin-walled sinus. The sinus possesses the highest automatic frequency of any portion of the heart and acts as pacemaker, and the wave of contraction passes successively over the auricle and ventricle. There is a marked delay of about 0·3 sec. at both the sino-auricular and auriculo-ventricular junctions (2). The rate of conduction of the wave of excitation is slow in the sinus and auricle and more rapid in the ventricle. The essential arrangement is that of a tubular heart with diverticula furnished with muscle of a specialised type. The sinus is characterised by high automaticity and low conductivity, whereas the ventricle is characterised by low automaticity and higher conductivity. The bulbus arteriosus has considerable automaticity, however, and if it is stimulated it may give rise to a series of 20–30 beats, each contraction being conducted backwards to the sinus venosus.

ELASMOBRANCHII

The hearts of Elasmobranchs are similar to those of the Cyclostomes, except that the specialisation of the different portions is greater, and there is a greater difference in the rate of conduction in the different chambers. The Elasmobranchs show a well-developed bulbus arteriosus which contracts rhythmically like the rest of the heart. The Elasmobranch heart is a powerful and efficient organ, and in *Scyllium* the blood pressure in the gill arteries is about 30 mm. Hg, and in the abdominal artery about 10 mm. Hg (Schönlein and Willem (5)).

TELEOSTEI

The chief difference between the hearts of Teleosts and those of Elasmobranchs is that in the former the bulbus arteriosus is rudimentary, and connects with a well-developed bulbus aorta. This is composed of plain muscle, does not contract rhythmically, and is morphologically and functionally a part of the blood vessels and not of the heart.

DIPNOI

In the Dipnoi an important modification appears, namely a partial separation between the oxygenated blood returning from the lungs and the blood returning from the rest of the body. Boas has given a full description of the heart of *Ceratodus*. The lung veins open into a separate portion of the sinus venosus and from this the blood passes into the auricle. The auricle is divided by an incomplete septum into two divisions; the right contains purely venous blood and the left a mixture of venous and oxygenated blood.

The auricular septum projects into the ventricle and partly divides this, and the long spiral bulbus aortae is divided by a spiral valve. This arrangement results in the oxygenated blood from the left side of the ventricle passing to the first two pairs of gill arteries and to the head, while the purely venous blood from the right side of the heart passes to the third and fourth gill arteries.

AMPHIBIA

In the Amphibia the heart consists of a sinus venosus, two auricles, a ventricle and a bulbus arteriosus. The bulbus arteriosus is composed of cardiac muscle and is contractile. The sinus venosus is the pacemaker for the whole heart. The pulmonary veins open into the left auricle, while the blood from the rest of the body is collected into the sinus venosus which opens into the right auricle. The bulbus arteriosus contains a complex spiral valve which ensures that the best oxygenated blood passes to the head and the least oxygenated blood to the lungs.

The frequency of the frog's heart is about 30 per minute at 15° C., and the systolic blood pressure is about 40 mm. Hg.

REPTILIA

In the Reptiles the division of the heart into a right and left side is carried a stage further than in the Amphibia, for the ventricle is partly divided in most genera, and in the Crocodilia there is a complete ventricular septum. No bulbus arteriosus is present in the Reptile's heart which consists of a sinus, two auricles and two ventricles.

The frequency of the reptilian heart varies from 10 to 60 per minute, and the blood pressure from 40–100 mm. Hg. Existing Reptiles vary over a much greater range of size than existing Amphibia, and are capable of more rapid movements and more prolonged exertion, and this difference is probably due in part to the superior efficiency of their circulatory mechanism.

CORONARY CIRCULATION IN COLD-BLOODED VERTEBRATES

The vertebrate heart requires a rich supply of oxygen in order to perform its work. In the Fishes, coronary arteries run back from the first efferent branchial arteries and these supply the heart with oxygenated blood.

In the Amphibia the ventricle is of a sponge-like structure, and the heart cells obtain their oxygen from the blood in the heart. Minute coronary arteries branch out in the bulbus arteriosus and supply this portion of the heart.

In the Reptiles the ventricles are supplied by blood from the left aorta.

HEART VALVES IN COLD-BLOODED VERTEBRATES

The vertebrate heart is provided with three sets of valves: (1) sino-auricular, (2) auriculo-ventricular, and (3) aortic valves.

The sino-auricular opening is provided with a pair of valves in all the cold-blooded Vertebrates.

The auriculo-ventricular opening also is provided with a pair of valves in the Fishes and Amphibia, but in the Reptiles the ventricular septum divides these valves and there is a pair of valves on each side. In the Crocodiles there is in addition on the right side a third valve formed out of the ventricular muscle.

The aortic valves vary greatly in the different classes. In the Elasmobranchs the long bulbus arteriosus is furnished with several rows of valves, but in the Teleosts the bulbus arteriosus is small and there is only one row of valves.

The Dipnoi and the Amphibians are provided with complex spiral valves in the bulbus arteriosus. Finally, in the Reptiles there is a single row of valves at the origin of the aorta and pulmonary artery.

LYMPHATIC HEARTS

In certain Fishes and Amphibia rhythmically contractile lymphatic hearts occur which circulate the lymph. The eels alone among Fishes possess such hearts, which are situated near the end of the tail.

Polimanti (3) has described the lymph heart of a large conger. The heart consisted of two chambers one on each side of the body. One acted as an auricle and collected lymph; it contracted first and emptied into the other chamber, the ventricle, and this, when it contracted, emptied the lymph into the vena caudalis. The frequency of the heart varied from 15 to 180 per minute. The activity of the heart depended on the integrity of the spinal cord, and when this was destroyed the heart was arrested.

The frog possesses two pairs of lymph hearts—an anterior pair between the transverse processes of the 3rd and 4th vertebrae, and a posterior pair lying one on each side of the posterior end of the urostyle.

These latter hearts are easy to observe and their functions have been carefully analysed. They are composed of a thin layer of striated muscle the cells of which do not anastomose, and which appear to be of a type intermediate between cardiac and skeletal muscle. The hearts contract rhythmically from 40 to 60 times a minute. They possess however no automatic rhythmicity, for their contractions usually cease instantaneously when the spinal cord is destroyed.

Direct stimulation of the lymph hearts can produce an extra systole, but this is not followed by a compensatory pause. The heart muscle resembles ordinary striped muscle, since there is no refractory period, it can be tetanised (Priestley (4)) and the force of contraction is proportional to the stimulus (Brücke (1)). The lymph heart is paralysed by curare (Priestley) and by potassium (Straub (6)). In their general behaviour these lymph hearts show a considerable resemblance to the haemal hearts of *Limulus*. They are interesting in that their properties offer such a striking contrast to the properties of the vertebrate haemal heart.

References

(1) Brücke. *Pflüger's Arch.* **115**, 33. 1916.
(2) Carlson. *Zeit. f. all. Physiol.* **4**, 259. 1904.
(3) Polimanti. *Zeit. f. Biol.* **59**, 171. 1913.
(4) Priestley. *Journ. of Physiol.* **1**, 19. 1878.
(5) Schönlein and Willem. *Zeit. f. Biol.* **32**, 511. 1895.
(6) Straub. *Arch. f. exp. Path. und Pharm.* **85**, 123. 1919.

Chapter V

GENERAL PHYSIOLOGY OF HEARTS OF WARM-BLOODED VERTEBRATES

General Structure—Heart Valves—Embryonic Heart—Oxygen Consumption—Need for Oxygen.

GENERAL STRUCTURE

The distinctive character of the circulation in warm-blooded animals is the complete separation of the pulmonary and systemic circulations. The pressure required to drive the blood through the pulmonary capillaries is much less than that needed for the systemic circulation, and in Man for example the pressure in the

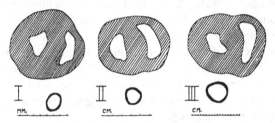

Fig. 6. Transverse sections of ventricle and aorta of I. Rat; II. Sheep; III. Horse. The sections have been enlarged or reduced photographically to bring them to the same size. The scales indicate the natural sizes.

pulmonary artery is only about one-fifth of that in the aorta. The right ventricle in consequence has much less work to do than the left ventricle, and the wall of the former is only about one-third as thick as that of the latter. The general structure and proportions of the heart are remarkably constant in both Birds and Mammals of all sizes. This is exemplified in Fig. 6 which shows sections of the hearts of the rat and the horse reduced to an equal measure. In both Birds and Mammals the sinus venosus is not present as a separate chamber, but is represented by an area near the junction of the superior vena cava on the right auricle. This area acts as pacemaker to the heart. There is no bulbus arteriosus, and the heart consists of four chambers. A special feature of the hearts of warm-blooded animals is the presence of a specialised conductive tissue which transmits the wave of excitation from the auricle to the ventricle. This will be described in chapter VII.

The remarkable uniformity in the hearts of warm-blooded animals is probably due to the fact that the maximum possible efficiency of circulation is a most urgent need in all these animals, and hence there is little possibility for non-significant variations in the heart.

The contrast between the work of the heart in warm-blooded and cold-blooded Vertebrates is shown by the following calculation.

The heart of a frog of 25 grm. at 15° C. has a frequency of 30 per minute; the output of blood is about 2 c.c. a minute, and the blood pressure is about 40 mm. Hg.

The heart of a bird or a mouse of this weight has a frequency of 700–1000 per minute; the output of blood is about 60 c.c. per minute from the two ventricles together, and the systemic blood pressure is about 70 mm. Hg.

These figures indicate the enormous difference between the performance of the hearts of cold-blooded and warm-blooded animals of the same size; the heart of the small bird or Mammal must do about 70 times as much work per gram muscle per minute as does the heart of the frog.

HEART VALVES IN BIRDS AND MAMMALS

The valves which guard the sino-auricular opening in cold-blooded Vertebrates are reduced in Birds and Mammals to a small fold between the pacemaker and the rest of the right auricle.

In Birds the left auriculo-ventricular opening is guarded by membranous valves, but the right auriculo-ventricular opening is provided with a thick muscular strap which closes the orifice when it contracts. This arrangement is remarkable since functionally it is entirely different from any of the other valves found in vertebrate hearts.

In Mammals the auriculo-ventricular valves are all membranous; on the left is the mitral valve with two flaps and on the right the tricuspid valve. The orifice of the aorta and pulmonary arteries are each guarded by a row of three valves in Birds and Mammals.

EMBRYONIC HEARTS IN VERTEBRATES

The embryology of the chick's heart has been worked out more exactly than that of any other Vertebrate.

The heart commences as a simple tube of cells forming a syncytium, and this tube commences a spontaneous beat before any trace of nervous tissue appears. His found that automatic rhythm was shown by the chick's heart on the 2nd day of incubation, and that nerve fibres only reached the heart on the 5th day.

Electrocardiographic records have been taken of the chick's embryo heart by several workers (2, 7, 12, 15).

Külbs (7), and Spadolini and Giorgio (12) agree that the three waves P, R, and T, can be distinguished at the age of 50 hours. At this time the rate is 60–70 per minute, the P–R interval 0·1 sec., and the duration of the R and T waves combined is 0·4 sec. Convolution and constriction of the heart do not commence until later than 50 hours, and therefore the functional division of the auricular from the vestibular portion of the heart begins before the morphological division. At this stage Fano (5) described the auricular portion of the heart as contracting spontaneously and stated that the auricular contraction was followed by a peristaltic wave which passed down the ventricle. The rate of conduction in the ventricle he found to be 36–115 mm. per second. He also found that if the heart were divided, the isolated auricular portion contracted from 80 to 140 times per minute, whereas the isolated vestibular portion contracted at only about one-third of this frequency. This last observation has been confirmed by Lewis (8).

At the end of the 4th day the R wave becomes much more pronounced and at this age the general character of the electrocardiograms resembles that of the adult animal. The pulse rate is 140 and the P–R interval 0·1 sec.

The electrocardiographic records and Fano's mechanical records together suggest that at first the heart is composed of a tube of muscle resembling plain muscle in its power of conduction of impulses, but that by the end of the 4th day the ventricular muscle has become capable of rapid conduction. Pickering (10) states that the vagal endings can be shown to be active at 200 hours.

Mammals

Little information is available concerning the physiology of the embryonic mammalian heart in its early stages. In the rabbit embryo pulsations have been observed on the 7th day (Bischoff) and in the human heart on the 18th day (Pflüger). During the latter stages of foetal life, when the four chambers of the heart are developed, there are various well-known modifications which adjust the foetal circulation to the placental supply of oxygen. Therefore, the activity of the foetal heart cannot be compared directly with that of the infant.

The rate of the foetal heart is slightly greater than that of the new-born animal. In Man the rate is about 135 and in cows about 150 (Ellinger).

THE OXYGEN CONSUMPTION OF THE HEART

The figures given on page 42 show that in the mammalian heart-lung preparation both the oxygen consumption and work done by the heart are dependent upon the initial filling of the heart. Hence the amount of oxygen consumed by a heart depends upon the conditions of the experiment. Evans and Starling (4) found that in the heart-lung preparation of the dog, working under conditions resembling those of bodily rest in the intact animal, the oxygen consumption was about 3–4 c.c. per gram per hour. Evans (3) calculated that under these conditions the efficiency of the heart was about 20 per cent.

The efficiency of the heart depends, however, upon the degree of filling as is shown in Table 8. The efficiency increases up to a certain maximum filling which probably is greater than that ever attained in life, as the filling of the heart *in situ* is limited by the rigid pericardium.

The figures for the oxygen consumption of the hearts of other vertebrates are less accurate than those obtained with the heart-lung preparation of a dog. Rohde (11) found that the isolated cat's heart utilised 2 to 4 c.c. oxygen per gram per hour. Weizsäcker (14) found that the isolated frog's heart working against a resistance of about 30 mm. Hg and with a frequency of about 25 per minute used 1·5 to 2·0 c.c. oxygen per gram per hour. He found that the resting metabolism of the frog's heart was about one-tenth of this figure.

The oxygen consumption of the Mammal's heart is similar to that of skeletal muscles during active work. For example, Barcroft and Kato (1) found that the oxygen consumption in the leg muscles of a dog *in situ* was 0·760 c.c. per gram per hour during rest and rose to 2·4 c.c. during work.

The figures for the oxygen consumption of the frog's heart are surprisingly high. Meyerhof (9) found that the oxygen consumption of the resting skeletal muscle of the frog at 14° C. was 0·014–0·028 c.c. per gram per hour.

THE NEED OF THE HEART FOR OXYGEN

The process of contraction in skeletal muscle has been shown to be anaerobic. Energy is provided by the breakdown of glycogen to lactic acid which is an anaerobic reaction. Oxygen is needed for the recovery process in which about one-fifth of the lactic acid is oxidised and the remainder reconverted into glycogen.

The skeletal muscle appears to be very tolerant of the accumulation of large quantities of lactic acid, and in consequence the body during a short period of violent exercise can incur a heavy oxygen debt.

Hill, Long and Lupton (6), for example, found that when a man did "standing running" for 4 minutes an oxygen debt of 18 litres was incurred and that the recovery process took 87 minutes. During this violent exertion the amount of oxygen used rose from 300 c.c. to 4000 c.c. per minute for a 73 kilo man. The amount of oxygen used during the 4 minutes' exercise was, however, in spite of this enormous increase in intake, less than the oxygen debt incurred.

Very little evidence is available concerning the chemical processes associated with the contraction of heart muscle, but the evidence available suggests that the processes are similar to those occurring in skeletal muscle. The heart muscle is, however, very rapidly depressed by any excess of acid, and it appears probable that it has much less power than skeletal muscle of incurring an oxygen debt by accumulating lactic acid. Consequently the heart is more dependent on a rich and continuous oxygen supply than is the skeletal muscle.

During violent exercise the work of the heart is increased very greatly. The total oxygen consumption of the body during violent exercise can increase 13-fold; it is true that both the oxygen utilisation from the blood and the mechanical efficiency of the heart are increased by increased filling, but the oxygen needed by the heart to maintain the work required of it during violent bodily exercise must be at least five times the normal.

With the isolated heart-lung preparation of the dog Evans and Starling obtained the following figures:

Table 2

	Blood flow c.c. per minute	Oxygen utilisation	Oxygen used c.c. per minute
Normal heart beat	62·5	0·32	3·75
Stimulated by adrenalin ...	83·0	0·75	12·5

Starling and Evans found in the same preparation that the normal blood supply of the heart was about 60 c.c. per minute per 100 grams heart muscle, but that during asphyxia this figure might rise to 370 c.c.

Tigerstedt[13] has collected the following figures for other organs of the dog:

Table 3

	Blood supply in c.c. per 100 grams organ per minute	
	At rest	At work
Skeletal muscle	13	85 (horse)
Stomach and gut	20–30	—
Submaxillary gland	68	168
Kidneys	50	100
Thyroids and suprarenals ...	500–600	—

Supposing that in the intact animal the oxygen utilisation coefficient of the coronary blood only rises to 0·6 which is double the normal value, then if the blood flow is increased two and a half times the oxygen supply will be raised 5-fold. The coronary flow in the dog can, however, increase 6-fold, and therefore in this case the coronary flow appears to be fully adequate to supply any quantity of oxygen needed by the heart, even when it is performing its maximum possible work.

References

(1) Barcroft and Kato. *Phil. Trans. Roy. Soc.* B, **207**, 149. 1915.
(2) Cluzet and Savronat. *Journ. de Physiol.* **16**, 802. 1916.
(3) Evans. *Recent Advances in Physiology*, p. 125. London, 1925.
(4) Evans and Starling. *Journ. of Physiol.* **46**, 413. 1914.
(5) Fano and Badano. *Arch. Ital. de Biol.* **13**, 387. 1920.
(6) Hill, Long and Lupton. *Proc. Roy. Soc.* B, **97**, 84. 1924.
(7) Külbs. *Beiträge. z. Physiol.* **1**, 439. 1920.
(8) Lewis. *Bull. Johns Hopkins Hosp.* **35**, 252. 1924.
(9) Meyerhof. *Pflüger's Arch.* **175**, 20. 1919.
(10) Pickering. *Journ. of Physiol.* **20**, 189. 1890.
(11) Rohde. *Arch. f. exp. Path. u. Pharm.* **68**, 401. 1912.
(12) Spadolini and Giorgio. *Arch. di Fis.* **19**, 479. 1921.
(13) Tigerstedt. *Physiol. der Kreislaufes*, **4**, 310. 1923.
(14) Weizsäcker. *Pflüger's Arch.* **68**, 401. 1912.
(15) Wertheim-Salomonson. *Pflüger's Arch.* **153**, 558. 1913.

Chapter VI

THE PROPERTIES OF VERTEBRATE HEART MUSCLE

Frog's Heart—Spontaneous Rhythmicity—Excitability—Conductivity—
"All or None" Response—Refractory Period—Time Relations—Comparison with other forms of Muscle—Invertebrate Heart Muscle—Vertebrate Heart Muscle—Initial Filling and Work.

FROG'S HEART

The characteristic properties of the cardiac muscle have been analysed more fully in the frog than in any other Vertebrate and therefore it is convenient to make the frog's heart muscle a basis of comparison. The chief distinctive properties of the heart muscle of the frog, *Rana temporaria*, may be summarised as follows:

SPONTANEOUS RHYTHMICITY

The spontaneous rhythmicity of the heart muscle varies in the different chambers of the heart. The sinus venosus has a natural frequency of about 30 per minute at 15° C. This rhythmicity is only slightly affected by pressure, and the sinus will continue its rhythmic contractions when there is no pressure upon it. The spontaneous rhythmicity of the auricle is feebler and an excised auricle under no tension usually does not beat, but a pressure of one or two centimetres of water usually produces a rhythm of about 12 per minute. The excised ventricle under no pressure does not beat spontaneously but a rhythm can be induced by increasing the pressure.

The following figures are given by Ludwig and Lucksinger[10]:

Table 4

Pressure in cm. water	Frequency per minute		
	Sinus	Auricle	Ventricle
0	28	0	0
5	—	18	—
20	39	40	30

These figures show clearly the relation between pressure and frequency, a relation which is seen still more clearly in the molluscan heart. Increase of pressure also produces increase in the force and amplitude of the frog's heart and this effect will be discussed later.

The frequency of all portions of the heart is increased when the temperature is increased, and the exact relation between frequency and temperature is discussed on page 64.

Finally there is an intimate relation between the size of animals, the metabolic rate, and the frequency of the sinus, which is discussed in chapter XI.

EXCITABILITY

The determination of the excitability is difficult in a muscle which executes spontaneous rhythmic contractions, as the excitability varies according to the position in the cycle at which the stimulus is introduced. Most determinations of excitability have been made on the frog's ventricle arrested by a Stannius ligature.

Various facts have however been established regarding the nature of the response of this tissue to electrical stimulation. The chronaxie of a tissue is determined by ascertaining the minimal current strength which can produce excitation when the duration of stimulus is indefinitely long, and then measuring the minimal duration of stimulus which produces excitation when twice the minimal current strength is applied.

The chronaxie of the ventricle depends on the nature of the electrodes, but with pore electrodes the chronaxie of the ventricle is about 5σ (Adrian (2)), whereas the chronaxie of skeletal muscle stimulated with pore electrodes is about $0\cdot3\ \sigma$ (Jinnaka and Azuma (5)).

Lucas (9) found that when two stimuli, 5 per cent. below threshold, were applied they produced summation in the ventricle at a maximal interval of 8σ, whereas when similar stimuli were applied to the frog's sartorius the maximal interval was $1\cdot5\ \sigma$.

These facts indicate that the local excitatory process produced by stimulation proceeds more slowly in the heart muscle than in skeletal muscle.

CONDUCTIVITY

Any stimulus capable of producing a measurable response in a piece of heart muscle is transmitted throughout the muscle. The speed of its transmission in the auricle or ventricle of the frog is about 100 mm. per second. Moreover the stimulus is conducted from chamber to chamber of the heart, although there is a delay of about 0·5 second at the junction between any two of the four chambers of the heart.

The wave of excitation can be conducted in any direction through the heart, and it is now generally agreed that the conduction is myogenic and not neurogenic.

The functional efficiency of any heart composed of more than one chamber is dependent upon the mode of conduction of the wave of excitation, for on this depends the accurate timing of the contractions of the different chambers.

In all such hearts it is essential that there should be sufficient delay between the contractions of the different chambers to permit the passage of the blood from one chamber into the next. Hence the mode of transmission of the wave of contraction is relatively complex, and will be considered in detail in the next chapter.

"ALL OR NONE" RESPONSE

The "all or none" response of the heart to stimuli is one of the most characteristic properties of vertebrate heart muscle. This effect appears to be due to the fact that the heart cells form a syncytium and hence the wave of excitation is transmitted freely in all directions through the heart muscle. Any stimulus sufficient to excite a single cell produces a disturbance which spreads to all the cells.

THE REFRACTORY PERIOD

A heart cell, once it has received a stimulus, is incapable of responding to a second stimulus for a period which is known as the absolute refractory period. The absolute refractory period ends about the end of the T wave of the electrical response and is followed by a relative refractory period during which time the excitability of the muscle is less than normal.

The most important consequence of the occurrence of this refractory period is that it is impossible to produce tetanus in normal cardiac muscle.

If cardiac muscle is stimulated artificially and the frequency of the stimulus is gradually increased, a point is reached when the interval between the stimuli is shorter than the refractory period and then the heart only responds to alternate stimuli. Another consequence is that no summation of effective stimuli is possible, that is to say, if the heart responds to a stimulus, the introduction of a second stimulus during the response does not affect the amplitude or duration of the response. The refractory period is followed by a supernormal phase during which the heart gives a response of greater amplitude than normal (Adrian (1)).

TIME RELATIONS OF RESPONSE OF HEART MUSCLE

Stimulation of the ventricle of the frog causes an immediate electrical response, and a mechanical response which appears after a measurable latent period. This latent period however becomes much shorter when special methods are adopted to eliminate mechanical factors producing delay and the true latent period of the mechanical response is probably not more than about 0·01 sec. The duration of the electrical response at any one point is about 0·7 sec. and that of the mechanical response is about 0·9 sec. These figures vary with a number of factors but seldom fall outside the limits of 0·5 and 1·5 sec.

COMPARISON BETWEEN CARDIAC AND OTHER FORMS OF MUSCLE

A much greater knowledge than we at present possess of the nature of muscle contraction is needed to understand fully the significance of the special properties of heart muscle.

The significance of some of these properties seems however clear. Plain muscle consists of long thin spindle-shaped cells with no direct connection between the cells. The simplest forms of cardiac muscle, e.g. the cells of the sinus venosus of the frog, resemble plain muscle closely in their general form but they are striated and the cells are connected by thin strands. In the frog's ventricle the cells are shorter and thicker, the striations clearer and the connections between the cells thicker. In the mammalian ventricle these changes are carried a stage further. Skeletal muscle appears to be a further development from plain muscle, in that numbers of cells have fused completely to form fibres, and these fibres have no automaticity and are separated so completely that excitation is not conducted from fibre to fibre. The skeletal muscle is in fact specialised so as to perform accurate rapid movements completely under the control of the central nervous system.

As regards the automaticity of the heart, this is a property it shares with most undifferentiated protoplasm, and the skeletal muscle must be regarded as a specialised form of tissue that has lost a characteristic common in undifferentiated cells. The conductivity of the wave of excitation in the heart is much more rapid than in plain muscle, a fact easily explicable on account of the heart cells forming a syncytium. On the other hand the complete fusion of cells that occurs in skeletal muscle fibres makes the conduction within a single fibre much more rapid than in the heart, but the fibres being separated there is no direct conduction from fibre to fibre.

The "all or none" law appears to be a general law regulating all forms of muscle. The work of Lucas(8), Pratt(11) and other workers has shown that the "all or none" law applies to the individual muscle fibre in skeletal muscle, and that the graded response of skeletal muscle is due to variation in the number of fibres excited.

In plain muscle conduction is very poor and it seems probable that the graded response observed to varying stimuli is due to the tendency of a wave of excitation to die out, and therefore the larger the number of cells excited at first the further the excitation spreads.

In the heart conduction is universal and hence as a general rule any stimulus sufficient to excite enough cells to produce a measurable stimulus produces a wave of excitation which spreads all over the muscle.

The refractory period appears to be a genuine peculiarity of cardiac muscle. So far as is known a muscle cell of either plain or skeletal muscle can maintain a state of contraction for a considerable time, whereas a heart cell cannot do so. This peculiarity is a very obvious safeguard for an organ upon the continuous rhythmic activity of which the life of the individual depends.

The refractory period of the heart muscle is a very different phenomenon from any observed in either skeletal or plain muscle. The refractory period in skeletal muscle is only about 2σ, which is similar to the length of the refractory period in nerves, hence the muscle can respond as fast as it can receive stimuli from the nerves, and since the refractory period is far shorter than the time required for the development of maximum tension, tetanus can

Table 5

Properties of frog's muscle (at 15° C.)

	Plain muscle (stomach)	Cardiac muscle (auricle or ventricle)	Skeletal muscle
Automaticity	Spontaneous rhythm 1–3 per min.	Spontaneous rhythm 10–20 per min. (aur.)	None
Conduction	5–20 mm. per sec.	100 mm. per sec. (aur. and vent.)	4000 mm. per sec.
Refractory period	None for contraction, but conduction is impaired after stimulation	Well marked for both contraction and conduction, and lasts for duration of contraction	Very short $(1-2\sigma)$, and ends long before contraction is completed
Summation of effective stimuli	Well marked	None	Well marked
Relation of response to intensity of stimulus	Response increases with increasing intensity of stimuli and tetanus can be produced	"All or none" response	Response increases with increasing intensity of stimulus and tetanus can be produced
Effect of increasing resting length of fibre	Frequency and amplitude of response increased	Frequency and amplitude of response increased	Amplitude of response increased
Chronaxie (Lapicque(7))	1–3 sec.	0·005 sec.	0·0003 sec.
Duration of isotonic response	60 sec.	0·5–1·0 sec.	0·1 sec.

be easily produced. In plain muscle the refractory period is probably longer, but is much shorter than the time required for the development of maximum tension, and here also tetanus can readily be produced. In vertebrate cardiac muscle, under normal conditions the absolute refractory period is considerably longer than the time required to produce maximum tension, and hence it is impossible to produce tetanus.

The preceding table summarises the outstanding differences between plain, cardiac and skeletal muscle in the frog. The properties of plain muscle vary very greatly and the stomach muscle has been taken as an example.

INVERTEBRATE HEART MUSCLE

The characteristics of invertebrate heart muscle have been described in previous chapters. It was there shown that the arthropod heart muscle resembled vertebrate skeletal muscle more nearly than vertebrate cardiac muscle.

The hearts of Worms, Molluscs and Tunicates however resemble vertebrate plain muscle more nearly than vertebrate heart muscle, but the heart muscle of the higher Molluscs approaches vertebrate cardiac muscle in some of its qualities.

VERTEBRATE HEART MUSCLE

There appears to be no marked difference between the fundamental properties of the frog's heart muscle and the properties of the heart muscles of other cold-blooded Vertebrates. The hearts of Cyclostomes, Fishes, Amphibia and Reptiles all show the same fundamental characteristics.

The outstanding difference between the heart muscles of the cold-blooded and the warm-blooded Vertebrates is one of speed. The heart of the mouse beats about twenty times as fast as that of a frog, and the a.-v. interval in the mouse is less than one-tenth as long as that of the frog.

Similarly the rate of conduction in the dog's ventricle is about ten times as great as the rate of conduction in the frog's ventricle. This difference can however partly be accounted for by the difference in the body temperature. The curves in Fig. 8 indicate that if the frog's heart could function at $37°$ C. its frequency at this temperature would be similar to that of the rabbit's heart.

The fundamental properties of the heart muscle of warm-blooded animals appears to be very similar to those of cold-blooded Vertebrates, but the specialisation of the functions of the tissues in different portions of the heart is carried further in the former.

THE RELATION BETWEEN THE FILLING OF THE HEART AND THE WORK PERFORMED

A general rule has been found to hold good for all forms of muscle investigated, namely, that the work performed in response to a stimulus depends on the initial length of the muscle fibre.

In the case of the skeletal muscle of the frog the tension and also the heat produced on stimulation both increase when the initial extension is increased up to a certain maximum, beyond which there is a decline in both.

The same relation has been shown to exist between the initial extension and the tension produced in the retractor penis of the dog, an example of mammalian plain muscle (Winton (12)).

In the case of the heart the fibres run in all directions and their exact length cannot be determined, but their average length must vary approximately as the circumference of the heart and therefore as $\sqrt[3]{\text{contents}}$.

The relation between the volume of the frog's heart in diastole and the force exerted in systole was established by Frank (1895), who showed that the force of the isometric contraction and the extent of the isotonic contraction were both dependent upon the initial volume.

The following figures by Kosawa (6) show the relation in the case of the tortoise's heart.

Table 6

Initial volume of fluid in heart in c.c.	0·5	1·0	1·5	2·0	2·5	3·0
Maximum pressure during isometric contraction in mm. Hg	15	35	45	47	47	48

Similarly Kosawa showed that the amplitude of the isotonic contraction, which is shown by the volume output of the heart, was proportional to the initial filling which is proportional to the venous pressure.

Table 7

Venous pressure in cm. water	2	10	16	21·6
Output of heart per min. in c.c.	1·3	9·4	14·7	19

This same law has been demonstrated equally clearly in the mammalian heart-lung preparation. The following are typical figures obtained by Evans and Matsuoka [4] in the heart-lung preparation of the dog and recalculated by Evans [3]. In this case the factor varied was the amount of blood admitted per beat into the right auricle.

Table 8

Volume output in litres per hour	Work per hour kg.m.	O_2 used per hour c.c.	Mechanical efficiency %
17	21	140	7
34	48	159	14·5
81	176	300	28

These figures show very clearly how the amount of venous filling regulates the amplitude of the heart's contraction, the work done, the oxygen consumed and also the mechanical efficiency of the heart.

Accurate measurements correlating the force of contraction with the initial tension in the heart, or with the initial volume of the heart, are only available in the case of the vertebrate heart.

Numerous observations on the hearts of Invertebrates all agree that both the frequency and amplitude of contraction in the invertebrate heart are dependent upon the degree of filling of the heart.

This is indeed one of the most obvious phenomena noted when working with invertebrate hearts. This relation between the venous pressure and the output of the heart provides a very simple mechanism whereby the work done by the heart varies according to the amount of blood supplied.

This is probably the most important mechanism for adjusting the circulation in Invertebrates. The vertebrate heart is subject to a complex and complete system of nervous control and therefore in this case the venous filling is a factor of rather less relative importance. Starling has however shown that even in the Mammal's heart the influence of venous pressure on cardiac output is of great practical importance.

References

(1) Adrian. *Journ. of Physiol.* **54**, 1. 1920.
(2) Adrian. *Journ. of Physiol.* **59**; *Proc.* lxii. 1924.
(3) Evans. *Recent Advances in Physiology*, p. 122. London, 1925.
(4) Evans and Matsuoka. *Journ. of Physiol.* **49**, 378. 1915.
(5) Jinnaka and Azuma. *Proc. Roy. Soc.* B, **94**, 77. 1922.
(6) Kosawa. *Journ. of Physiol.* **49**, 233. 1915.
(7) Lapicque. *C. R. Soc. de Biol.* **67**, 280 and 283. 1909.
(8) Lucas. *Journ. of Physiol.* **33**, 125. 1905.
(9) Lucas. *Journ. of Physiol.* **39**, 461. 1909.
(10) Ludwig and Lucksinger. *Pflüger's Arch.* **25**, 211. 1881.
(11) Pratt. *Amer. Journ. of Physiol.* **44**, 517. 1917.
(12) Winton. *Journ. of Physiol.* **61**, 386. 1926.

Chapter VII

THE TRANSMISSION OF THE EXCITATORY PROCESS

Tubular Hearts—Invertebrate Hearts—Conduction in Cold-blooded Vertebrates—Conduction in Birds' Hearts—Conduction in Mammalian Hearts—Electrical Response of Vertebrate Hearts.

TUBULAR HEARTS

The heart may be regarded as developing from a tube of muscle cells forming a syncytium. From this tube are developed chambers surrounded by specially differentiated muscle. Simple tubular hearts are found in worms, caterpillars, Tunicates and embryos, and in all these the rate of conduction is relatively slow, as is shown in the following table:

Table 9

Animal	Temperature °C.	Rate of conduction in mm. per second	Author
Lumbricus	15	20	Clark [6]
Nereis	13	10–20	Clark [6]
Lucanus larva	Room	19–44	Lasch [20]
Cossus cossus larva	15	20–40	Seliškar [32]
Ciona	Room	20–35	Carlson [5]
Phallusia	15	10–20	Clark [6]
Chick embryo	39	36–115	Fano and Badano [12]

INVERTEBRATE HEARTS

The arthropod hearts appear to be divisible into two chief classes. In the first class the contraction passes as a slowly conducted peristaltic wave. This class includes the insect larvae, many adult Insects and *Limulus* embryo and the rate of conduction as is shown in the table is about 20–40 mm. per second. In the second class the whole heart contracts almost simultaneously. In the case of *Limulus* the conduction is almost certainly neurogenic and the rate is about 400 mm. per second. Many adult Insects and all the Crustacea have hearts of this type, but the rate of conduction in these is not known, nor is it known whether the conduction is myogenic or neurogenic. In the Mollusca the heart is differentiated into auricles and a ventricle, and it is of interest to note that there is an interval of about 0·5 sec. between the contractions of the auricle and ventricle. There is no evidence available concerning the rate of conduction in the auricles or ventricles of Mollusca.

CONDUCTION IN COLD-BLOODED VERTEBRATES

In the cold-blooded Vertebrates the following figures are available for the rate of conduction of the wave of contraction down the heart.

Table 10

Rates of conduction in hearts of cold-blooded Vertebrates

	s.-a. interval in sec.	Conduction in auricle mm. per sec.	a.-v. interval in sec.	Conduction in ventricle mm. per sec.
Petromyzon (Carlson (4))	0·25	3–4	—	25
Dogfish (Straub (36))	—	—	0·6	—
Eel (Roskam (30) and Bakker (1))	0·4	—	0·38	129
Goldfish (Lewis (22))	—	—	0·2	—
Triton (Spadolini (34))	0·25	—	0·22	—
Rana temp.	0·18 (Skramlik (33)) 0·4 (Clark (6))	40–177 (Engelmann (10))	0·3	—
Bufo terrestris	—	—	0·3	100 (Lewis (23))
Tortoise	—	100 (Seliškar (32))	0·6	100–150 (Lewis (23)) 100–300 (Fredericq (15))

There are no satisfactory figures available concerning the conduction in the sinus venosus. Kupelwieser (19) found that the rate of conduction in the snake's sinus was about 168 mm. per second. Observations by the author suggested that the rate of conduction in the frog's sinus was about 100 mm. per second.

A comparison of the figures in Table 10 shows that conduction in all parts of the heart of the lamprey is much slower than in the hearts of other Vertebrates.

The rate of conduction in the sinus, auricle and ventricle of the other cold-blooded Vertebrates appears however to be in all cases between 100 and 200 mm. per second.

In all cases there is a marked delay at the sino-auricular and auriculo-ventricular junction.

For example, in the frog the impulse passes over the sinus at a rate of about 100 mm. per second, and then halts for 0·2 to 0·4 sec. at the sino-auricular junction, the conduction passes over the auricle at about 100 mm. per second and then halts about 0·2 sec. at the auriculo-ventricular junction, and passes on to the ventricle where the conduction rate is over 100 mm. per second, finally there is another halt of about 0·2 sec. between the ventricle and bulbus.

The mode in which the delay at the sino-auricular and auriculo-ventricular junctions is produced is of considerable interest. The delay is obviously essential to the working of the heart, since it allows time for the blood impelled from one chamber to enter the succeeding one before this begins to contract.

Skramlik (33) has shown that in the sinus, auricle and ventricle of the frog the long spindle-shaped muscle fibres are arranged in bundles which branch in all directions forming a network.

At the s.-a. and a.-v. junctions, however, he found that all the fibres were arranged in a circular ring. He accounted for the delay in conduction at these points by supposing that the rate of conduction is inversely proportional to the number of cells per unit distance that the wave of excitation has to traverse. This number will obviously be far the greatest at the s.-a. and a.-v. junction, where all the cells are arranged transversely to the direction in which the wave of excitation passes.

The manner in which the wave of excitation is conveyed from auricle to ventricle varies considerably in the different cold-blooded Vertebrates. In Fishes the whole of the junctional ring between the auricle and ventricle appears to conduct (Roskam (30)), in Amphibia only about three-quarters of the ring conducts (Nakano (27)), and in Reptiles only about one-half of the ring conducts (Laurens (21)).

Mackenzie (25) found that in the Fish's heart the s.-a. junction consisted of a simple ring of tissue, while the a.-v. junction was formed by a portion of the auricle (the auricular canal) which was invaginated into the a.-v. orifice. In the Reptiles he found that the s.-a. junctional tissue was reduced to a single bundle of fibres and that the auricular canal was continuous with the ventricular muscle on the left side of the heart. The paths of conduction from auricle to ventricle in vertebrate hearts are represented diagrammatically in Fig. 7.

One of the first stages in the process of development of the heart appears therefore to be the division of the primitive tube by specialised rings of muscle fibres arranged transversely as sphincters. In the Reptiles these rings are reduced to localised strands of muscle connecting the adjacent chambers. This process of specialisation of conductive tissue is carried considerably further in the hearts of Birds and Mammals.

Fig. 7. Conduction in auricular funnel in vertebrates. I. Fishes. II. Amphibians. III. Reptiles. IV. Birds. V. Mammals. (After Mangold.)

In the frog the tissue connecting the auricles and ventricles possesses certain functional peculiarities, for the power of this tissue to conduct impulses is lower than that of the auricle or ventricle. Eckstein (8) showed that either the auricle or ventricle could be driven at a rate more rapid than the a.-v. connection could transmit impulses, and that either an a.-v. or a v.-a. block could be induced by stimulating the auricle or ventricle respectively at a sufficiently rapid rate. It has been repeatedly shown that the rate of the a.-v. conduction is greater than that of the v.-a. conduction, and this indicates some specialised path of conduction. The a.-v. conduction, moreover, is more readily influenced by drugs or by changes in ionic concentration than is the activity of either the auricle or ventricle.

Skramlik's explanation of the delay at the s.-a. and a.-v. junction has been mentioned, but there are other possible explanations.

There is a considerable amount of connective tissue at the a.-v. junction in all cold-blooded Invertebrates and the number of muscle fibres connecting the auricle with the ventricle is relatively small. Engelmann (9, 11) showed that if a heart of a cold-blooded animal was cut into thin strips, then the conduction in the strips might be as little as 30 mm. per second, which is about one-third the normal rate, and he also found that such thin strips showed an abnormally long refractory period.

Gaskell (16) showed that if a tortoise's auricle or ventricle was cut so as to leave two halves divided by a thin bridge, then there was a considerable delay of conduction at the bridge. Gaskell also showed that partial section of the a.-v. connection in the tortoise prolonged the a.-v. interval considerably.

The connection between auricle and ventricle in cold-blooded animals appears therefore to consist of muscle cells so arranged as to give a considerable delay to the passage of conduction. In the Fishes the ring is uniform, but in the Amphibia the power of conduction varies in different portions of the ring, whilst in the Reptile only certain portions of the ring can conduct.

The delay in conduction can be explained by the mode of arrangement of the fibres or by the narrowing down of the possible paths of conduction, but also the presence of a special form of junctional muscle fibre is possible.

CONDUCTION IN THE HEARTS OF BIRDS

In the hearts of Birds the sinus is fused with the auricle and the following tissues are present, the sino-auricular node or pace-maker, the auricles, the auriculo-ventricular conductive tissue and the ventricles.

The rate of conduction in the auricles is about 2000 mm. per second (Mangold and Kato (26)), and the average rate of conduction in the auriculo-ventricular conductive tissue and ventricles is about 2000–4000 mm. per second (Lewis (23)).

The a.-v. interval (i.e. interval between P and R waves in electrical response) is from 0·06–0·1 sec. (hen's heart, Lewis (23), Spadolini and Giorgio (35); pigeon's heart, Lewis (23), Kahn (18); goose's heart, Firket (13); swan's heart, Lewis (23)).

In the Bird's heart large specialised fibres (Purkinje fibres) are present. These are very well developed in the auricle of the ostrich but do not appear in the hen's auricle (Holmes (17)). They are much scarcer in the ventricle and do not form any definite con-necting strand between auricle and ventricle. Their significance for the conduction of excitation is therefore obscure. Flack (14) concluded that in birds the a.-v. conducting tissue was localised in small patches scattered around the a.-v. ring. Mangold and Kato (26) found that functionally the conduction of excitation from auricle to ventricle was localised in a single portion of the a.-v. ring in the interventricular septum at the base of the valve of the right auriculo-ventricular orifice. They found that this tract was not composed of any form of histologically specialised tissue.

There appears therefore to be a specialised channel for the conduction of excitation from the auricles to the right ventricle in birds, but no special channel for conduction to the left ventricle has been discovered.

CONDUCTION IN THE HEARTS OF MAMMALS

In the mammalian heart the normal impulse originates in the sino-auricular node, which is situated near the junction of the superior vena cava with the right auricle. The wave of excitation spreads from here over both auricles; it is however transmitted to the ventricle by a single channel composed of tissue with functional and morphological peculiarities.

This channel commences at the auriculo-ventricular node which lies in the interauricular septum and passes from here by the bundle of His. This leads down to the interventricular septum, and then divides into right and left branches, which split up into fibres and these pass to all portions of the two ventricles.

The auriculo-ventricular node is composed of small muscle cells and nerves, while the rest of the conducting tissue is composed of Purkinje cells, which are much longer than the ordinary cells of heart muscle, and have less marked striations than the latter.

The rate of conduction in the various systems in the dog has been studied exhaustively by Lewis [24], who gives the following figures:

Table 11

	Rate of conduction in mm. per sec.	Time spent in passage sec.
Auricle	800	0·03
a.-v. node	200 or less	0·05
Purkinje fibres ...	4000	0·01–0·015
Ventricular muscle ...	400	0·01

Total P–R interval 0·09–0·1 sec.

These figures show that conduction along the specialised path of conduction, namely the bundle of His and its branches, is very

c

4

rapid. This system is composed of Purkinje cells, interlaced with a nerve plexus. Conduction in the hearts of warm-blooded Vertebrates is believed to be purely myogenic and the Purkinje fibres are believed to represent a form of heart cell specialised for rapid conduction. It is however a remarkable fact that in the Birds' hearts conduction is almost if not quite as rapid as in the Mammals' hearts, but in the former the Purkinje fibres do not apparently form a path of conduction, even though these fibres occur in large numbers, in some birds such as the ostrich, scattered throughout the auricles and ventricles.

There are no figures for the rates of conduction of the wave of excitation in the hearts of other Mammals comparable in exactitude with those obtained by Lewis in the dog. These latter figures show however that the conduction in the dog's heart is about ten times as rapid as that in the frog's heart.

In the Mammals, as in all other Vertebrates, there is a considerable delay owing to the conduction being slowed at the junctional tissue between the auricle and ventricle, and this delay at the auriculo-ventricular node accounts for at least half of the observed P–R interval.

Comparatively few writers have measured the P–R interval in any but the common domestic animals and man, but the author has collected the figures shown in the table below. Many of these figures have been calculated by the writer from electrocardiographs reproduced in original papers.

Table 12

Animal	P–R interval in sec.	Author
Vesperugo pipistrellis	0·02–0·03	Buchanan [2]
Dormouse	0·04	Buchanan [3]
Mouse	{ 0·025	Clark [6]
	{ 0·025	Oppenheimer [28]
Rat	{ 0·054	Clark [6]
	{ 0·037	Oppenheimer [28]
Guinea-pig	{ 0·05	Pütter [29]
	{ 0·03–0·045	Schott [31]
Cat	0·08	Lewis [23]
Rabbit	0·03	Schott [31]
Dog	0·1	Lewis [23]
Man	0·15	Lewis [22]
Horse	0·3	Kahn [18] and Waller [37]
Elephant	0·3	Colin, Forbes and Campbell [7]

The most striking thing about these figures is that the *P–R* interval varies so little in different animals.

The *P–R* interval in the rat for example is one-sixth that of the horse, and since the length of the rat's ventricle is only about one-twentieth that of the horse's ventricle, a much greater proportion of the *P–R* interval of the horse must be spent in conduction of the impulse over the auricles and ventricles than in the rat, unless indeed the rate of conduction in the rat's auricles and ventricles is far lower than in the horse's, a very improbable supposition. The delay at the a.-v. junction therefore varies relatively little in different species of Mammals.

ELECTRICAL RESPONSE OF VERTEBRATE HEARTS

The study of the mode of transmission of the wave of excitation over the heart has been facilitated greatly by the use of the string galvanometer, which permits the measurement of the time at which any excitation arrives at any point within about 1 σ.

The exact relation of the electrical to the mechanical response of the heart is unknown, but it is important to remember that the string galvanometer can measure a difference of potential at two points of a small fraction of a millivolt, and that the amount of energy needed to produce such a difference is an insignificant fraction of the energy released in the contraction of the heart. The electrical record is therefore of great value for ascertaining time relations, but tells us little regarding the force of contraction of the heart.

References

(1) Bakker. *Zeit. f. Biol.* **59**, 335. 1913.
(2) Buchanan. *Journ. of Physiol.* **40**; *Proc.* xliii. 1910.
(3) Buchanan. *Journ. of Physiol.* **42**; *Proc.* xxi. 1911.
(4) Carlson. *Zeit. f. all. Physiol.* **4**, 208. 1905.
(5) Carlson. *Amer. Journ. of Physiol.* **16**, 67. 1906.
(6) Clark. Unpublished observations.
(7) Colin, Forbes and Campbell. *Amer. Journ. of Physiol.* **55**, 385. 1921.
(8) Eckstein. *Pflüger's Arch.* **157**, 541. 1914.
(9) Engelmann. *Pflüger's Arch.* **11**, 465. 1875.
(10) Engelmann. *Pflüger's Arch.* **56**, 188. 1895.
(11) Engelmann. *Pflüger's Arch.* **62**, 449. 1896.
(12) Fano and Badano. *Arch. Ital. de Biol.* **13**, 387. 1890.

(13) Firket. *Arch. Intern. de Physiol.* **12**, 22. 1912.
(14) Flack. *Arch. Intern. de Physiol.* **12**, 120. 1912.
(15) Fredericq. *Comp. Rend. Soc. de Biol.* **85**, 239. 1921.
(16) Gaskell. *Journ. of Physiol.* **4**, 64. 1883.
(17) Holmes. *Journ. of Physiol.* **58**; *Proc.* cxi. 1923.
(18) Kahn. *Pflüger's Arch.* **162**, 67. 1915.
(19) Kupelwieser. *Pflüger's Arch.* **182**, 50. 1920.
(20) Lasch. *Zeit. f. all. Physiol.* **14**, 312. 1913.
(21) Laurens. *Pflüger's Arch.* **150**, 139. 1913.
(22) Lewis. *Clinical Electrocardiography.* London, 1913.
(23) Lewis. *Phil. Trans. Roy. Soc.* B, **207**, 221. 1916.
(24) Lewis. *Proc. Roy. Soc.* B, **89**, 560. 1917.
(25) Mackenzie.   17th *Intern. Congress of Med.* (*Anat. and Embr.*), p. 123. 1913.
(26) Mangold and Kato. *Pflüger's Arch.* **160**, 91. 1914.
(27) Nakano. *Pflüger's Arch.* **154**, 373. 1913.
(28) Oppenheimer. *Zeit. f. d. ges. exp. Med.* **28**, 96. 1922.
(29) Pütter. *Pflüger's Arch.* **168**, 209. 1917.
(30) Roskam. *Arch. Intern. de Physiol.* **15**, 141. 1919.
(31) Schott. *Arch. f. exp. Path. u. Pharm.* **87**, 309. 1920.
(32) Seliškar. *Journ. of Physiol.* **61**, 172. 1926.
(33) Skramlik. *Zeit. f. d. ges. exp. Med.* **14**, 246. 1921.
(34) Spadolini. *Arch. di Fis.* **17**, 233. 1919.
(35) Spadolini and Giorgio. *Arch. di Fis.* **19**, 495. 1921.
(36) Straub. *Zeit. f. Biol.* **42**, 363. 1901.
(37) Waller. *Journ. of Physiol.* **47**; *Proc.* xxxii. 1913.

Chapter VIII

THE NERVOUS CONTROL OF THE HEART

Nature of Conduction—Nervous Control in Cold-blooded Vertebrates—
Birds—Mammals—Variations in Vagal Control in Birds and Mammals—
Embryonic Hearts—Action of certain Drugs on Hearts.

NATURE OF CONDUCTION

In previous chapters it has been shown that the frequency and force of the heart is partly regulated by the degree of diastolic filling which modifies the initial length of heart fibres.

In all classes of animals the activity of the heart is further regulated by nervous control, in nearly all cases augmentor and depressor nerves are present.

For many years there has been a controversy regarding the extent of the nervous control of the heart, but certain outstanding facts regarding the limits of nervous control appear to be fairly well settled.

The classical experiments of Engelmann and Gaskell established a very strong probability that the initiation and conduction of the wave of excitation in the vertebrate heart was purely myogenic. This view was strengthened by the observation of His jun. and other workers, who showed that the embryonic chick heart commenced to beat on the 2nd day of incubation, but that no nerves reached the heart until the 5th day of incubation. The frequency of chick hearts on the 2nd day is about 200 per min.

Tissue-culture experiments have provided additional proof, for Burrows [5] noted that cultures of embryo chick hearts which contained no trace of nerve cells contracted rhythmically. Fischer [10] has shown that when two cultures of chick's heart are grown close together the cultures beat with independent rhythms until they coalesce, after which the whole pulsates synchronously. This shows that in this tissue the conduction is myogenic, and is strong evidence in favour of the view that the origin and conduction of the wave of excitation in all Vertebrates is myogenic in character. There is a strong probability that this is also the case in the worms and Molluscs and in larval forms of Arthropods. On the other hand, Carlson (7) has provided strong evidence that the origin and conduction of the beat in *Limulus* heart is neurogenic in character. The evidence is inconclusive regarding other Arthropods.

The lymph hearts of the frog are another example of a rhythmically contractile circulatory organ in which the origin of the beat is purely neurogenic.

In all hearts, however, the activity of every cell is to a greater or less extent controlled and regulated by augmentor or depressor nerves or both, hence all hearts are richly supplied by nerve plexuses, and nerves can usually be traced to every cell.

NERVOUS CONTROL IN COLD-BLOODED VERTEBRATES

Cyclostomes and Fishes

The nervous control of the lampreys appears to be feeble. Green (15) failed to find any evidence of heart-regulating nerves in *Petromyzon*. Carlson (6) found that the vagus had no action on the heart of the larvae of *Petromyzon entosphenus*, but that in adult *Petromyzon* accelerator and probably inhibitor nerves were present. Zwaardemaker (48) found that in *Petromyzon fluviatiles* both accelerator and inhibitory nerves were present.

In all Fishes other than the lampreys the vagus nerves have a powerful action. Stimulation of the vagus produces arrest of the heart both in Elasmobranchs and Teleosts. Section of the vagi causes acceleration of the heart, and this proves that the vagus normally exerts a tonic depressor action on the heart.

Brücke (3), when summarising the action of the vagus in fishes, states that in all recorded cases stimulation of the vagus arrests the heart but never causes any preliminary slowing or diminution in the force of the beat.

MacWilliam (29) however showed that the eel's heart can be slowed by reflex stimulation of the vagus. MacWilliam found that vagal stimulation on the eel diminished the force of contraction of the auricle but had no certain action on the force of contraction of the ventricle.

Queen (38) found that in teleost hearts vagal stimulation could arrest the heart but did not produce any diminution in the force of contraction of the auricle or ventricle.

Brücke (3) states that there is no evidence for the existence of accelerator nerves in the Fish's heart.

Nervous control of Amphibian Hearts

Amphibian hearts are under the control of depressor and augmentor nerves.

There is a marked seasonal variation in the action of the vagus in the frog, for stimulation produces much less effect on the heart in the summer than in the winter. The vagus controls the activity of all cells in the frog's heart. Stimulation of the vagus produces the following effects: diminution in frequency, reduction in force of beats of auricle and ventricle, and prolongation of the s.-a. and a.-v. interval. Hoffmann (19) found that vagal stimulation also reduced the rate of conduction in the auricle and ventricle.

Several investigators have tried to determine if the vagus exercises any tonic control over the heart of the frog, but there is no clear evidence that section of the vagus causes any increase in rate.

Stimulation of the sympathetic nerves causes an increase in the frequency and the force of contraction of both auricle and ventricle in the frog (Skramlik (40)).

Nervous control of Reptile Hearts

According to Gaskell (12) the tortoise differs from the frog in that the vagus while producing a depressor action on the sinus, auricle and a.-v. connective tissue yet exerts no direct action on the ventricle. François Frank (11) however states that the vagus produces a direct depressor effect on the tortoise's heart. In other Reptiles the vagus has a powerful action on the ventricle. Vagal stimulation depresses the response of the ventricle more than of the auricle in Lacerta viridis (MacWilliam (30)). In the water tortoise (Pseudemys), the snake and the alligator the vagus acts more powerfully on the auricle than on the ventricle (Mills (33, 33a and 33b)). The augmentor sympathetic nerves act on all chambers of the heart in the alligator (Mills (33a)).

ACTION OF VAGUS ON AVIAN HEART

The action of the vagus on Birds' hearts is on the whole less powerful than on other vertebrate hearts. Usually only an arrest of a few seconds can be produced. Ducks are an exception, since in these vagal stimulation can produce an arrest of some minutes.

Stimulation of the vagus causes a diminished force of contraction of auricles (Knoll(24)) and of ventricles (Jurgens(23)). The vagi exercise a tonic influence on Birds' hearts; Stübel(43) studied the effect of section of the vagi on a large number of Birds and found that in all cases section caused acceleration of the pulse rate, although this effect was much more marked in some species than in others.

ACTION OF AUGMENTOR NERVES ON AVIAN HEART

Stübel found that stimulation of the 2nd and 3rd dorsal nerves caused slight increase in frequency and an increase in the force of contraction of the auricle. No effect was observed on the ventricle.

ACTION OF VAGUS ON MAMMALIAN HEART

Vagal stimulation produces slowing or arrest of the heart, diminution in force of muscular contraction, and prolongation of a.-v. interval.

The vagus has a particularly powerful depressor action on sino-auricular and auriculo-ventricular junctions. In the auricle vagal stimulation produces depression of force of beat (many authors, including Wiggers(47)). The refractory period of the auricle is reduced but the rate of conduction in the auricle (Lewis, Drury and Bulger(26)) is not reduced.

The extent of the action of the vagus on the mammalian ventricle has been the subject of much controversy. Tigerstedt(45) concludes that probably the vagus has a direct action on the frequency and force of the ventricular contraction, but that the vagal action is much weaker on the ventricle than on the auricle.

ACTION OF AUGMENTOR NERVES ON
MAMMALIAN HEART

Augmentor fibres pass to the sympathetic chain in the white rami communicantes of the 2nd and 3rd dorsal nerves, and to a lesser extent in those of the 1st and 4th dorsal nerves. The cell stations of these nerves are situated partly in the 1st thoracic ganglion and partly in the inferior cervical ganglion. Stimulation of the augmentor nerves produces an increase of frequency which in the dog's heart may amount to more than 50 per cent. The augmentor nerves also cause an increase in the force of contraction of both auricle and ventricle (Roy and Adami, 1872) and the a.-v. interval is shortened (Bayliss and Starling).

The augmentor nerves exercise only a slight tonic influence on the mammalian heart. For example, Hunt[21] found in the dog that the normal heart rate was 100, and after section of the accelerans the rate was unaltered, on the other hand when the vagus was frozen and the accelerans intact the rate was 154, whilst when both nerves were paralysed the rate was 123.

SUMMARY OF THE NERVOUS CONTROL OF
THE HEART OF VERTEBRATES

Augmentor nerves apparently do not occur in Fishes, but in all other Vertebrates these nerves appear to act on all the heart chambers. The vagus nerve acts as a cardiac inhibitor in all Vertebrates with the possible exception of the lampreys, but the extent of their control of the different chambers of the heart varies considerably. In all cases the vagal stimulation has a more powerful action on the sinus venosus than on any other portion of the heart. In all cases except a few Reptiles the vagus has a greater action on the auricle than on the ventricle.

The action of the vagus on the ventricle varies considerably in that it has no action on the Fish's ventricle, only a feeble action on the ventricle in most Reptiles, in Birds and in Mammals, but has a powerful action on the ventricle in Amphibia.

DIFFERENCES IN THE DEGREE OF VAGAL CONTROL OF THE HEART IN WARM-BLOODED ANIMALS

The degree of vagal control in Mammals and Birds varies very greatly.

(1) Animals with small hearts (i.e. ratio $\frac{\text{heart wt.} \times 100}{\text{body weight}}$ less than 0·4) have rapid pulses for their size when at rest, and section of the vagus causes little acceleration.

(2) Animals with large hearts (i.e. ratio more than 0·6) have slow pulses for their size when at rest, and section of the vagus causes great acceleration. The following are some striking examples of this difference.

Table 13

Animal	Small-hearted animals		
	Ratio $\frac{\text{heart wt.} \times 100}{\text{body weight}}$	Normal rate of heart	Heart rate after vagal section
Rabbit (17)	0·27	205	321
Hen (42)	0·5	288	312

Animal	Large-hearted animals		
	Ratio $\frac{\text{heart wt.} \times 100}{\text{body weight}}$	Normal rate of heart	Heart rate after vagal section
Hare (31)	0·75	64	264
Pigeon (42)	1·5	120	300

Stübel (42) noted that in all Birds with large hearts the vagus had a powerful action (e.g. pigeon, duck, sea-gull and hawk), on the other hand there was little vagal control in the hen, rook and jackdaw. In animals in which the vagus exercised a considerable tonic influence, stimulation of the vagus also produced a powerful inhibitory action on the heart. The vagus appears to exercise little tonic control on the hearts of small animals, for Oppenheimer (35) found that the pulse rates of small Birds and Mammals was unaffected by large doses of atropine, as is shown in the following table:

Table 14

Animal	Body weight in grm.	Pulse rate before drug	Drug	Pulse rate after drug
Mouse	16	670	Atropine 1 mg.	670
	—	655	Adrenaline 0·02 mg.	700
Bramblefinch ...	23	900	Atropine 0·05 mg.	694
	—	694	Adrenaline 0·02 mg.	729
Sparrow ...	18	442	Atropine 0·1 mg.	764
Blackbird ...	55	400	Adrenaline 0·02 mg.	360
	—	360	Atropine 0·5 mg.	630

As regards the action of atropine, the figures for the blackbird show that the dose was adequate to release the vagus. The normal pulse rate for a warm-blooded animal of about 20 grm. is 600–1000 and when the pulse was at this rate atropine produced no effect.

In the case of the sparrow the pulse rate before the atropine was given was abnormally slow and in this case atropine produced acceleration.

Harrington (16) found that vagal section had no effect on the pulse rate of the guinea-pig, and the author found the same to be true in the rat. The author also found that large doses of atropine produced no acceleration of the pulse rate of either the mouse or the rat. Stübel (42) found that in small Birds vagal section produced less acceleration than in large Birds.

In small animals the high metabolic rate appears to necessitate the animal maintaining as a normal pulse rate a frequency which is almost the maximum which the heart can maintain, and therefore there is no vagal control normally exercised. Oppenheimer's observations also show that adrenaline does not accelerate the pulse rate in small animals, and this also suggests that the normal rate is very near the maximum possible rate.

There are some other examples of lack of vagal control which are unexplained. Gley and Quinquard (14) found that the pulse rate of the goat was unaffected by vagal section. In man vagal control appears to be abolished in typhoid fever, for although the pulse is relatively slow in this disease, yet atropine produces no acceleration.

VARIATIONS OF THE VAGAL CONTROL WITH AGE

In man it is well known that tonic vagal control is greatest in early adult life and is less at the extremes of life. Many observers have shown that in new-born animals vagal section produces no increase in heart rate (Anrep [1] in kittens; Meyer [32] in puppies).

Lhotak von Lhota [27] measured the response to atropine of puppies of various ages and obtained such figures as these:

Age of puppy in days	...	30	210
Normal pulse rate ...	...	240	106
Pulse rate after atropine	...	246	260

NERVOUS CONTROL IN EMBRYONIC HEARTS

In chick embryos Botazzi [2] found that stimulation of the vagal nerves had no certain action on the 18th–20th day, but that it had a well-marked action on the new-born chick. Lewis [25] found that sympathetic fibres reached the heart on the 5th day.

Pickering [37] found however that after 160 hours, direct stimulation of the chick embryo heart produced inhibition and that the same effect was produced at this time by muscarin. Before this age neither of these procedures had any inhibitor effect. Pickering found that in rat embryos, even at the earliest stages, muscarin produced inhibition of the heart and that this action was antagonised by atropine.

THE ACTION OF CERTAIN DRUGS ON THE HEART

In the Vertebrates the administration of adrenaline produces a reaction which is almost identical with the effect of sympathetic stimulation.

The parallel is not absolutely exact; for example, sympathetic stimulation causes secretion of sweat, whereas this effect is not produced by adrenaline.

Similarly administration of pilocarpine and physostigmine produce almost exactly the same effects as stimulation of the vagus and other parasympathetic nerves, and the action of these drugs and also the action of the parasympathetic nerves is antagonised by atropine.

These facts led to the assumption that these drugs acted on nerve endings and hence it was assumed that, if adrenaline or pilocarpine produced an action on an organ, this was evidence that the organ was supplied by sympathetic or parasympathetic nerves respectively.

This latter assumption is obviously a dangerous one, for in some mammalian tissues irregular actions have been observed. For example, pilocarpine stimulates the uterus of many species of Mammals, but there is no evidence that any motor parasympathetic nerves run to the uterus.

The work of Loewi[28] and his pupils has however thrown additional light on this subject. Loewi has shown that in an isolated frog's heart stimulation of the vagus nerve releases into the perfusion fluid an inhibitory substance, which can be demonstrated by transferring the fluid to another heart. He also found that stimulation of the sympathetic nerves liberates an augmentor substance in the heart.

Loewi finally has shown that administration of atropine does not prevent the liberation of inhibitor substance on vagal stimulation, but that this substance cannot produce an action on the atropinised heart cell. This work suggests that adrenaline, pilocarpine and atropine do not act on nerve endings but on the muscle cells, and that stimulation of the appropriate nerves liberates substances with an action resembling adrenaline and pilocarpine, and that the action of these substances on the muscle cells can be antagonised by ergotoxine and atropine respectively.

Consequently even if adrenaline or pilocarpine produces an action on a tissue and the action is antagonised by ergotoxine or atropine respectively, this is not a conclusive proof that the tissue is supplied by either sympathetic or parasympathetic cells. Even in the Vertebrates for example the rectus abdominis of the frog is stimulated by acetyl-choline and this action is antagonised by atropine, but there is no evidence that the muscle is supplied by parasympathetic nerves.

It is however of interest to note the facts observed regarding the action of these drugs on various hearts.

Action of Adrenaline on Invertebrate Hearts

Adrenaline has been shown to produce acceleration of rhythm and augmentation of the force of beat in the hearts of nearly all Invertebrates. Such actions of adrenaline have been described in the contractile vessels of leeches (Gaskell[13]), in the hearts of various Molluscs, *Aplysia* (Heymans[18]), *Pecten* (Hogben and

Hobson (20)) and in the hearts of Arthropods, *Limulus* heart muscle and heart ganglion (Carlson (8)), heart of *Maja* (Hogben and Hobson (20)), heart of lobster (Brücke and Satake (4)). Seliškar found that adrenaline increased the rate of conduction in the heart of the larva of *Cossus cossus*. Elliott (9) stated that adrenaline did not stimulate the heart of the crab, but this appears to require confirmation.

Adrenaline also acts on other invertebrate tissues than the heart. It stimulates the intestine of the crayfish (Ten Cate (44)), the oesophagus of *Aphrodite*, *Aplysia* (Hogben and Hobson (20)) and of *Helix* (Ten Cate). It is interesting to note that in Vertebrates the inhibitory actions of adrenaline are as common as its augmentor actions, but that in Invertebrates only augmentor actions have been described.

The Action of Muscarin on Invertebrate Hearts

Muscarin produces a depressant action on most of the invertebrate hearts that have been studied.

In *Aplysia* (Straub (41) and Heymans (18)) and in *Octopus* (Ransom (39)), muscarin arrests the heart and this action is not antagonised by atropine. In *Helix* (Vulpian (46)) muscarin inhibits the heart but the action is antagonised by atropine.

In *Limulus* (Nukada (34)) muscarin arrests the heart and the action is antagonised by atropine, but in *Daphnia* (Pickering (36)) muscarin has no action on the heart.

Hunter (22) found that muscarin had a depressant action on the heart of *Salpa*. This corrects an old observation of Kruckenberg, who found that the drug had no action on this heart.

Muscarin appears therefore to exercise very generally an inhibitory action on invertebrate hearts, but the antagonistic action of atropine is sometimes absent.

References

(1) Anrep. *Pflüger's Arch.* **21**, 78. 1880.
(2) Botazzi (1897). Quoted from Tigerstedt. *Phys. d. Kreisl.* **2**, 321. 1921.
(3) Brücke. Winterstein's *Hand. d. vergl. Phys.* **1**, 1018. 1923.
(4) Brücke and Satake. *Zeit. f. all. Physiol.* **14**, 28. 1912.
(5) Burrows. *Münch. med. Woch.* **59**, 1473. 1912.
(6) Carlson. *Zeit. f. all. Physiol.* **4**, 259. 1904.

(7) Carlson. *Amer. Journ. of Physiol.* 16, 230. 1906.
(8) Carlson. *Ergebn. d. Physiol.* 8, 447. 1907.
(9) Elliott. *Journ. of Physiol.* 32, 401. 1905.
(10) Fischer. *Journ. of Exp. Med.* 39, 577. 1924.
(11) François Frank. *Arch. de Physiol. Norm. et Path.* 5th ser. 3, 475. 1891.
(12) Gaskell. *Journ. of Physiol.* 4, 89. 1883.
(13) Gaskell. *Journ. of Gen. Physiol.* 2, 73. 1919.
(14) Gley and Quinquard. *Arch. Néerl. de Physiol.* 7, 392. 1922.
(15) Green. *Amer. Journ. of Physiol.* 6, 318. 1902.
(16) Harrington. *Amer. Journ. of Physiol.* 1, 385. 1898.
(17) Hering. *Pflüger's Arch.* 60, 429. 1895.
(18) Heymans. *Arch. Intern. de Pharm.* 28, 337. 1923.
(19) Hoffmann. *Zeit. f. Biol.* 67, 434. 1917.
(20) Hogben and Hobson. *Brit. Journ. of Exp. Biol.* 1, 487. 1923.
(21) Hunt. *Amer. Journ. of Physiol.* 2, 426. 1897.
(22) Hunter. *Amer. Journ. of Physiol.* 10, 1. 1904.
(23) Jurgens. *Pflüger's Arch.* 129, 516. 1909.
(24) Knoll. *Pflüger's Arch.* 67, 591, 595. 1897.
(25) Lewis. *Bull. Johns Hopkins Hosp.* 35, 252. 1924.
(26) Lewis, Drury and Bulger. *Journ. of Physiol.* 54; *Proc.* xcvii, xcix. 1921.
(27) Lhotak von Lhota. *Pflüger's Arch.* 141, 515. 1911.
(28) Loewi. *Pflüger's Arch.* 189, 239; 193, 201. 1921.
(29) MacWilliam. *Journ. of Physiol.* 6, 192. 1885.
(30) MacWilliam. *Journ. of Physiol.* 6; *Proc.* xvi. 1885.
(31) MacWilliam. *Proc. Roy. Soc.* B, 53, 464. 1893.
(32) Meyer. *Arch. de Physiol.* 479. 1893.
(33) Mills. (*Pseudemys.*) *Journ. of Physiol.* 6, 251. 1885.
(33 a) Mills. (*Alligator.*) *Journ. of Anat. and Physiol.* 20, 550. 1886.
(33 b) Mills. (*Tropidonotus.*) *Journ. of Anat. and Physiol.* 22, 2. 1888.
(34) Nukada. *Mitt. a. d. Med. Fak. d. Kaiser. Univ. z. Tokyo*, 19, 1, 1917.
(35) Oppenheimer. *Zeit. f. d. ges. exp. Med.* 28, 96. 1922.
(36) Pickering. *Journ. of Physiol.* 17, 356. 1894.
(37) Pickering. *Journ. of Physiol.* 20, 165. 1896.
(38) Queen. *Zeit. f. Biol.* 62, 33. 1913.
(39) Ransom. *Journ. of Physiol.* 5, 261. 1884.
(40) Skramlik. *Zentralb. f. Physiol.* 34, 351. 1919.
(41) Straub. *Pflüger's Arch.* 103, 429. 1904.
(42) Stübel. *Pflüger's Arch.* 135, 249. 1911.
(43) Stübel. *Pflüger's Arch.* 135, 303, 313. 1911.
(44) Ten Cate. *Arch. Néerl. de Physiol.* 9, 172. 1924.
(45) Tigerstedt. *Physiol. d. Kreisl.* 2, 351. 1921.
(46) Vulpian. *Comp. Rend. Soc. de Biol.* 88, 1295. 1879.
(47) Wiggers. *Amer. Journ. of Physiol.* 42, 133. 1916.
(48) Zwaardemaker. *Arch. Néerl. de Physiol.* 9, 213. 1924.

Chapter IX

INFLUENCE OF TEMPERATURE ON HEART FUNCTIONS

Frequency and Metabolic Rate—Hibernating Animals—Other Functions of the Heart.

INFLUENCE ON FREQUENCY AND METABOLIC RATE

The influence of temperature on the frequency of rhythmic processes is extraordinarily uniform throughout the animal kingdom. There is an extensive literature on this subject and only a few typical examples will be considered.

Fig. 9 and Tables 15 and 16 show the effect of change of temperature on the rate of movement of *Amoebae*, the rate of movement of cilia of *Mytilus* and the frequency of the frog's heart; it will be seen that all three processes follow practically superimposable curves. Fig. 8 and Table 15 show the relation between the effect of change of temperature on the frequency of the isolated frog's heart and on the isolated rabbit's auricle and it will be seen that the two curves show a considerable resemblance.

Table 15

Temperature Coefficients ($Q/10$) of frequency and oxygen consumption in vertebrate tissues

Temp.	Frog's heart		Mammalian tissues			
	(i) Frequency (Clark (2))	(ii) Oxygen consumption of resting heart (Weizsäcker (16))	(i) Frequency of rabbit's auricles (Clark (2))	(ii) Frequency of dog's heart-lung preparation (Evans (4))	(iii) Oxygen consumption of dog's heart-lung preparation (Evans (4))	(iv) Oxygen consumption of isolated guinea-pig's uterus (Evans (5))
0–10°	3·5	—	—	—	—	—
5–15°	2·83	—	—	—	—	—
10–20°	2·4	2·5	—	—	—	—
15–25°	2·05	—	2·55	—	—	2·9
20–30°	1·83	2·5	2·17	—	—	2·51
25–35°	—	—	2·0	2·35	2·0	2·0
30–40°	—	—	2·07	1·8	—	1·7–1·9

A very large number of observations are available for very diverse tissues, and the general result is that the increase lies

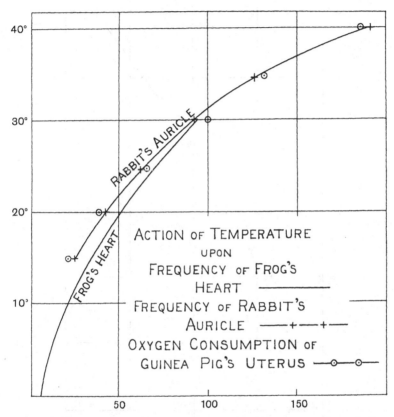

Fig. 8. Action of variation in temperature upon frequency and metabolism of isolated tissues.

Ordinate: temperature. Abscissa: frequency per minute and metabolic rate.

In the case of the metabolic rate the figures have been reduced to the scale: oxygen consumption at 30° C. = 100.

Figures for frequency by Clark (2). Figures for metabolism from Evans (5).

between two and threefold for a 10 degrees rise of temperature, within limits of about 10 degrees above or below the normal temperature of the animal; when the temperature is lowered more

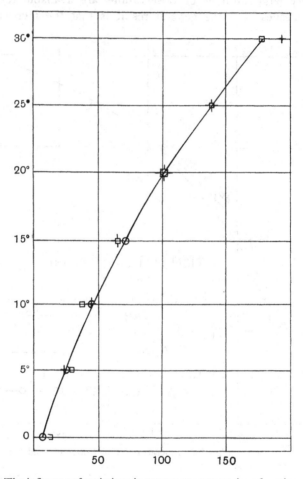

Fig. 9. The influence of variations in temperature on various functions of cold-blooded animals.

All functions reduced for comparison to the scale: rate at 20° C. = 100.

Ordinate: temperature. Abscissa: arbitrary scale.

      ☐ = frequency of isolated frog's heart (Clark (2)).

      ⊙ = rate of movement of amoebae (Pantin (12)).

      + = rate of ciliary movement in *Mytilus* (Gray (7)).

than about 10 degrees below the normal temperature of the animal, the alteration in activity produced by a change of temperature is increased, and a change of temperature of $10°$ C. produces a more than threefold change in the rate of activity ($Q/10 > 3$). On the other hand, when the temperature is raised above a certain point heat paralysis sets in and the activity decreases. The temperature at which heat paralysis commences differs in different tissues, and examples of such limits are $20°$ C. with *Amoebae*, about $33°$ C. with *Mytilus*, $35°$ C. with the tissues of *Rana temporaria*, and $45°$ C. with mammalian tissues.

Table 16

Temperature coefficients ($Q/10$) of processes in invertebrate tissues

Temp.	(i) Rate of movement of *Amoebae* (Pantin[12])	(ii) Rate of movement and oxygen consumption of cilia of *Mytilus* (Gray[7])
$0-10°$	7·33	3·1
$5-15°$	2·71	2·7
$10-20°$	2·17	2·3
$15-25°$	—	2·15
$20-30°$	—	1·95

Considerable attention has been given to the exact relationship between temperature changes and the frequency of the heart. The relation is not a simple linear one but in many cases follows van't Hoff's formula

$$K_1 = K_0 . \epsilon^{\frac{\mu (T_1 - T_0)}{2 \, T_1 \times T_0}}.$$

In this formula T_0 and T_1 are two temperatures reckoned from absolute zero; K_0 is the rate of reaction (or frequency in the case of the heart) at the temperature T_0, and K_1 is the rate at temperature T_1; μ is a constant.

The value of μ for most cellular processes lies between 11,000 and 16,000.

Many reactions follow this formula fairly closely but those processes, which are susceptible of the most accurate measurement, usually show a distinct deviation from the figure given by the formula, and therefore it is difficult to say what significance can be attached to the fact that the formula usually gives approximately accurate results.

The reason for the uniformity in the effect of alterations of temperature on the rhythm of the heart appears to be that the metabolic rate of all cells varies in a very uniform manner with the change of temperature, and the frequency of tissues contracting rhythmically varies directly as the metabolic rate.

Gray[7] found that the oxygen consumption and the rate of movement of cilia of *Mytilus* varied with change of temperature in exactly the same manner. Figure 8 shows that the frequency of the isolated rabbit's auricle and the oxygen consumption of the isolated guinea-pig's uterus vary in identical fashion. A comparison of the frequency of the hearts of mammals of different size shows that variations in frequency are closely correlated with variations in the metabolic rate as will be shown in Chapter x.

The frequency of the contraction of the frog's oesophagus (Stiles[13]) and the rabbit's gut (Fletcher and Alvarez[6] and Magnus[11]) vary with changes in temperature in a manner closely similar to the variations in frequency observed in the case of the heart.

HEARTS OF HIBERNATING ANIMALS

An extrapolation of the curve for the rate of the rabbit's auricle in Fig. 8 shows that the rates at 10° C. and 40° C. would be 14 and 190 respectively, which is a difference of about 14-fold; the curve also shows that the heart would cease to beat between 0° and 5° C.

The following heart rates have been found in hibernating animals:

Table 17

Animal	Hibernating Body temperature (° C.)	Rate	Awake Body temperature (° C.)	Rate	Author
Plecotus auritus	circa 10	76	circa 40	600–900	Buchanan[1]
Vesperugo pipistrellus	circa 10	30	—	660	Buchanan[1]
Hedgehog(*Erinaceus*)	circa 10	48	circa 40	280–320	Buchanan[1]
Marmot	11	38–43	38	160–206	Hecht[9]
Woodchuck and hedgehog (excised hearts)	0	2	40	over 120	Tait[14]

These figures show that the difference in rates in the hearts of animals hibernating and awake are of the order to be expected from the variations in heart rates in excised hearts of other animals. The metabolic rates of hibernating animals are reduced to an extent proportional to the reduction in their heart rates.

THE INFLUENCE OF TEMPERATURE ON VARIOUS FUNCTIONS OF THE HEART

The influence of temperature on the frequency of a rhythmically contractile tissue is particularly easy to measure.

A certain amount of information is available regarding the influence of temperature on other functions of the hearts of various animals.

A few typical figures are shown in Table 18; it appears that the

Table 18

Function	Tissue	Range of temperature $(T_0 - T_1)$	Alteration per 10° C. rise of temp.	Author
Rate of conduction of excitation	Medusa	20–30	2	Harvey and Meyer [8]
„	Ascidia atra	15–25	2·4	Hecht [9]
	—	25–35	1·9	
„	a.-v. junction frog's heart	15–35	1·5	Ishikama [10]
„	„	15–25	1·6	Clark [2]
Length of refractory period	Frog's ventricle	20–10	1·55	Trendelenburg [15]
„	„	28–3	1·4	Dennig [3]

rate of conduction is increased practically at the same rate as the frequency by rise of temperature, and that the rate of recovery, as indicated by the shortening of the refractory period, is influenced in a similar manner. The fact that these various functions of the heart are affected approximately equally is of considerable practical importance, for it is this fact that permits the heart to function over a considerable range of temperatures, as is seen in the case of hibernating animals. Most observers have noted in hibernating animals the occurrence of partial heart block between the auricles and ventricles. This fact suggests that the a.-v. conductive tissue is affected somewhat more by a fall in temperature than is the pacemaker of the heart. The manner in which the force of contraction of the heart is affected by change of temperature is somewhat complex. In the case of the frog, if the heart is allowed to beat at its natural frequency, then increase in temperature increases the work done per minute but decreases the force of contraction of the individual beats. If the sinus is removed and the heart driven with an artificial rhythm and the optimal frequency for each temperature is chosen, then the force of individual contractions increases as the temperature is lowered.

References

(1) Buchanan. *Journ. of Physiol.* **42**; *Proc.* xxi. 1911.
(2) Clark. *Journ. of Physiol.* **54**, 275. 1920.
(3) Dennig. *Zeit. f. Biol.* **72**, 187. 1920.
(4) Evans. *Journ. of Physiol.* **45**, 213. 1912.
(5) Evans. *Journ. of Physiol.* **58**, 22. 1923.
(6) Fletcher and Alvarez. *Amer. Journ. of Physiol.* **44**, 344. 1917.
(7) Gray. *Proc. Roy. Soc.* B, **95**, 6. 1923.
(8) Harvey and Meyer. *Publ. of Carnegie Inst. of Washington*, **132**, 22. 1911; **183**, 3. 1914.
(9) Hecht. *Zeit. f. d. ges. exp. Med.* **4**, 259. 1915.
(10) Ishikama. *Pflüger's Arch.* **202**, 308. 1924.
(11) Magnus. *Pflüger's Arch.* **102**, 143. 1904.
(12) Pantin. *Brit. Journ. of Exp. Biol.* **1**, 519. 1924.
(13) Stiles. *Amer. Journ. of Physiol.* **5**, 338. 1901.
(14) Tait. *Amer. Journ. of Physiol.* **59**, 467. 1922.
(15) Trendelenburg. *Arch. f. (Anat. u.) Phys.* 271. 1903.
(16) Weizsäcker. *Pflüger's Arch.* **148**, 535. 1912.

Chapter X

THE INFLUENCE OF BODY WEIGHT ON THE SIZE OF THE HEART IN WARM-BLOODED ANIMALS

Ratio between Body Weight and Heart Weight in Animals of same Species—Influence of Muscular Activity—Influence of Climate—Smallest and Largest Animals—Proportions of the Heart.

HEART AND BODY WEIGHT

The influence of body weight upon the functions of the heart can be shown most clearly in the Birds and Mammals, because the oxygen requirements of the body are far greater in warm-blooded than in cold-blooded animals, and hence the efficiency of the circulation is of much greater importance in the former.

This difference between cold-blooded and warm-blooded animals is illustrated by the fact that arrest of the circulation produces unconsciousness in the latter in a few seconds, whereas a cold-blooded Vertebrate, such as the frog, will continue to perform voluntary movements in an apparently normal manner for about half an hour after its heart has been arrested by a drug such as strophanthin. The clearest correlations between body weight and heart functions may therefore be expected to occur in Birds and Mammals, because in these classes the standard of efficiency of the circulation is the highest and hence there is the least chance of any extensive non-significant variations.

The Mammalia are remarkable for their extraordinary range of size. The smallest Mammal, the dwarf bat (*Vesperugo pipistrellus*), weighs about 3·5 grm. (Hesse [12]), whilst the largest whales attain weights of at least 100,000 kilos. This represents a 30,000,000-fold variation in weight. The Birds show a smaller range of weight. The smallest humming bird (*Mellisuga minima*) weighs 2 grm. [11], and a large ostrich weighs over 100 kilos. Larger birds existed in the past however and Pütter has estimated the weight of the extinct *Aepyornis* as 750 kilos.

These are the extreme limits of weight, but the commonly occurring weights show a smaller range. Excluding the bats, there are few species of Mammals which weigh less than 10 grm., and excluding the whales, there are only a few species which weigh more than 500 kilos. In the case of the Birds a considerable number weigh less than 10 grm. Heinroth(11) gives the weights of 436 species of Birds, and of these 23 species weigh less than 10 grm. The number of species weighing more than 5 kilos is however small, and nearly all of these do not fly.

The hearts of all these animals are of very similar general proportions and arrangement, and it is interesting that any single type of pump should be able to function efficiently over such an enormous range of sizes. It is therefore of interest to consider the manner in which the morphological and functional properties of the heart are adapted to meet the requirements of these varying sizes.

THE RELATION BETWEEN HEART WEIGHT AND BODY WEIGHT

The relation between the weight of the heart and the weight of the body is expressed conveniently by stating the heart weight as the percentage of the body weight, and this figure will be termed the heart ratio.

The heart ratios of about 150 species of Birds and 100 species of Mammals are plotted in Figs. 10 and 11 respectively. The majority of these figures were obtained from Hesse(12), or from the measurements shown in Appendix 2.

These figures show that there is a great variation in the heart ratios of animals of every weight, but that there is a definite general correlation between body weight and heart ratio.

The average figures for the heart ratio appear to follow the rule that heart weight varies as (body weight)$^{0.9}$. This relation gives the following relation between heart ratio and body weight:

$$\text{Heart ratio} = \frac{\text{heart weight} \times 100}{\text{body weight}}.$$

$$\text{If heart weight} = K\,(\text{body weight})^{0.9}.$$

$$\text{Then heart ratio} = \frac{K\,(\text{body weight})^{0.9} \times 100}{\text{body weight}}$$

$$= \frac{K \times 100}{(\text{body weight})^{0.1}}.$$

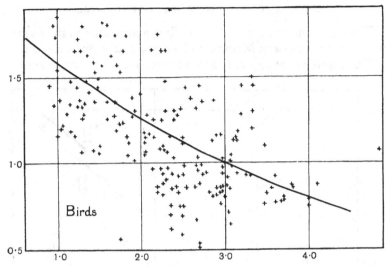

Fig. 10. Relation between heart ratio and body weight in birds.

Ordinate: heart ratio $\left(\text{i.e.} \dfrac{\text{heart weight} \times 100}{\text{body weight}} \right)$.

Abscissa: log. of body weight in grams.

The curve has been drawn to fit the equation:

$$\text{Heart ratio} = \frac{2\cdot0}{(\text{body weight})^{0\cdot1}}.$$

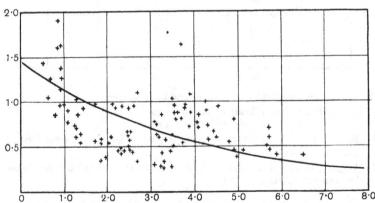

Fig. 11. Relation between heart ratio and body weight in mammals.

Ordinate: heart ratio $\left(\text{i.e.} \dfrac{\text{heart weight} \times 100}{\text{body weight}} \right)$.

Abscissa: log. of body weight in grams.

The curve has been drawn to fit the equation:

$$\text{Heart ratio} = \frac{1\cdot4}{(\text{body weight})^{0\cdot1}}.$$

The curves in Figs. 10 and 11 are drawn according to the above formula. In the case of Mammals $K = 0\cdot014$, and in Birds $K = 0\cdot02$. The rule does not appear to hold for Mammals over 100,000 grm.

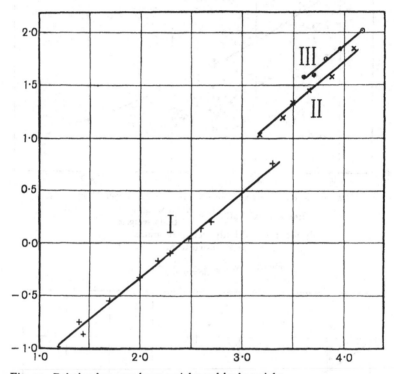

Fig. 12. Relation between heart weight and body weight.
  Ordinate: log. of heart weight in grams.
  Abscissa: log. of body weight in grams.

  I.   Rodents (tame mice, rats and rabbits). (Author's figures and Donaldson(5).)
  II. Infants (Müller(20)).
  III. Dogs. Averages of a total of about 150 figures collected by the author
       (references(14), (24), (25), (26)).

This rule is however only a rough generalisation, for inspection of Figs. 10 and 11 shows that the individual variations from the average are very great.

For example the ostrich (wt. 70 kilos) has a heart ratio of 1·08, which is as high as the ratio in many Birds weighing less than

50 grm. Again in the case of the Mammals the rat, the cat, the ox and the elephant all have heart ratios lying between 0·4 and 0·5 grm., and the mouse and the horse both have a heart ratio of about 0·7.

A clearer relation between body weight and heart weight is seen however if animals of related species and of similar mode of life are compared. Those domestic rodents, which normally rely on the protection of holes for safety, form such a group and for these the author obtained the following ratios:

Table 19

Animal	Weight	Heart ratio
Mouse	10	0·7
	25	0·65
Rat	100	0·48
	200	0·41
	500	0·32
Rabbit	2000	0·27

These figures which are plotted in Fig. 12 give a relation: heart weight varies as (body weight)$^{0·8}$, this is the same ratio that is found generally for adult animals of the same species but varying weights.

BODY WEIGHT AND HEART RATIO IN ANIMALS OF THE SAME SPECIES

A comparison of adult animals of the same species but of varying weights shows clearly an influence of body weight on the heart ratio. A number of such figures are shown collected in Table 20,

Table 20

	Small animals		Large animals	
	Body wt. in kilos	$\dfrac{\text{Heart wt.} \times 100}{\text{body wt.}}$	Body wt. in kilos	$\dfrac{\text{Heart wt.} \times 100}{\text{body wt.}}$
Frankonian sheep (Seeburger[12])	45·9	0·464	53·4	0·420
Southdown sheep (Seeburger[12])	37·03	0·452	46·8	0·429
Oxen (Schneider[12])	659	0·424	838	0·387
Bulls „	478	0·485	829	0·364
Cows „	445·3	0·509	530	0·468
Men (Gocke[12])	47	0·613	68	0·535
Dogs (Joseph[14] and others[24, 25, 26])	4·13	0·95	15·3	0·685
West American sparrow, Zonotrichus leucophrys (Dal Piaz[12])	0·02162	0·943	0·02594	0·888

most of these are quoted from Hesse's paper[12]. The figures for dogs were compiled from the sources quoted by the author.

The figures in Table 20 show that the heart weight varies as (body weight)$^{0.8}$ in the case of sheep, dogs and sparrows, but in the case of oxen, bulls, cows and men the relation in heart weight varies as (body weight)$^{0.6}$. This relation in the case of men appears doubtful as it does not agree well with estimations of heart size made during life by means of X-rays.

The area of the heart can be measured by X-rays and Diethlen[4] gives the following average figures for the relation between heart area and body weight in healthy men:

Table 21

| Body weight in kilos ... ... | 43 | 50 | 60 | 70 | 80 |
| Area of heart shadow in sq. cm. ... | 92 | 104 | 112 | 119 | 131 |

These figures show that the area of the heart varies as (body weight)$^{0.6}$ and this suggests that the volume of the heart must vary approximately as (body weight)$^{0.9}$, a figure which is nearer the figures found for other animals.

It is possible that the reason why the heart weight in cattle only increases as (body weight)$^{0.6}$ is that the extra weight in the heavier individuals is largely fat.

Figures for rats and humans of varying ages are available, and in rats (Donaldson[5]) the relation holds for body weights from 10 to 450 grm. that the heart weight varies as (body weight)$^{0.8}$: the same rule applies to infants during the first two years of life, but after this the increase in heart weight is almost directly proportional to the body weight (Müller[20]).

All these calculations only apply if average numbers are taken, for great individual variations may occur, for example the heart ratio in dogs of 5 kilos may vary from 0.73 to 1.2.

The relation between heart weight and body weight in animals of the same or of related species but of varying sizes and ages is shown in Fig. 12. These cases all follow the general rule that heart weight varies as (body weight)$^{0.8}$.

The relation between body weight and heart weight can be stated therefore as follows. When a comparison is made between animals of the same species but of varying ages and sizes, or when a comparison is made between animals of allied species and of a similar mode of life, then the relation is usually found that heart weight varies as (body weight)$^{0.8}$. When however a general comparison is made of animals of widely varying habits of life then this relationship is partly obscured, and the average relationship is approximately heart weight varies as (body weight)$^{0.9}$. The influence of body weight upon the heart ratio is much clearer in Birds than in Mammals, as is seen by a comparison of Figs. 10 and 11. The probable reason for this is that the majority of Birds are capable of flying considerable distances, and all of these correspond in their power of sustaining exertion to the athletic class of Mammals.

MODE OF LIFE AND HEART RATIOS

Table 22 shows the ten Mammals weighing more than 20 grm. which have the highest heart ratios, amongst the figures which I have collected, and the ten which have the lowest heart ratios. This table shows that as a general rule those animals which depend on speed either for the capture of food or for escape from enemies have exceptionally high heart ratios. The badger is the only exception, but this animal is capable of severe and continuous exercise in the form of digging. The animals with low heart ratios are those which cannot perform continuous and severe muscular exercise and many of them are animals which rely for safety upon holes.

Table 22

Animals with high heart ratios			Animals with low heart ratios		
Animal	Weight in grm.	Heart ratio	Animal	Weight in grm.	Heart ratio
Deer	20,600	1·15	Rabbit, wild	2,000	0·27
Badger	9,000	1·08	,, tame	1,600	0·3
Wolf	26,000	1·017	Field vole	68	0·37
Weasel	54	0·939	Hedgehog	886	0·38
Stoat	232	0·94	Wild boar	93,000	0·39
Fox	6,200	0·92	Norwegian rat	150	0·4
Seal	35,000	0·92	Sheep (domestic)	50,000	0·42
Red deer	300,000	0·90	Guinea-pig	400	0·42
Hare	3,000	0·77	Domestic cat	3,000	0·46
Dog	4,000	0·85	Domestic ox	500,000	0·49
,,	10,000	0·78			
,,	30,000	0·685			

Mammals can be divided roughly into three classes as regards the size of their hearts: (i) the smallest Mammals, all of which have high heart ratios; (ii) animals which are capable of continuous and severe muscular exercise, in whom the heart ratio is above 0·6; (iii) animals incapable of continuous and severe muscular exertion, in whom the heart ratio is below 0·6.

The qualification that exercise must be continuous is important, for animals with very small hearts are capable of violent exertion for a short time. For example the rabbit which has the lowest heart ratio of any Mammal can run fast for a short distance. The explanation of this is that short bursts of activity can be performed by running into debt for oxygen, but the efficiency of the circulation is the chief limiting factor for long-continued exertion. Small animals which undertake the relatively severe exertion of flight such as the bats, naturally have the highest heart ratios of any Mammals.

The relation between the heart ratio of Birds and their mode of life is more difficult to establish. For example the average heart ratio of nine species of Accipitres given by Hesse of weights between 100 and 1000 grm. is 0·94 and the average heart ratio of eleven species of Passeres of similar weights is the same. Loer[18] propounded the general rule that if the smallest Birds are left out of consideration then the highest heart ratios are found in those Birds which are the best flyers, the quickest runners or the loudest singers or callers.

THE INFLUENCE OF MUSCULAR ACTIVITY ON THE HEART WEIGHT

It has already been shown that powerful animals adapted for violent exercise have the highest heart ratios. The following are some striking examples of the difference in the heart ratios of active and inactive animals.

Table 23

Inactive animals			Active animals		
Animal	Body weight	Heart ratio	Animal	Body weight	Heart ratio
Barndoor cock[23]	1336	0·68	Game cock[23]	1663	1·03
Tame duck[3]	2060	0·63	Wild duck[3]	770	1·06
Tame rabbit	2000	0·24	Hare	4000	0·77

Schwarzenecker (quoted by Hesse[12]) states that the heart of a thoroughbred horse weighs 5·5 kilos as opposed to a weight of 4–4·5 kilos in an ordinary horse. Dogs bred for speed such as greyhounds have higher heart ratios than ordinary dogs.

There appears therefore no doubt that species or even breeds of animals distinguished for muscular power have unusually large hearts. There is however considerable dispute as to whether hypertrophy of the heart can be induced in individuals by severe muscular exercise.

Külbs[15] took two puppies from the same litter and made one work and allowed the other a tranquil existence. At the end of the period he found that the heart ratio in the working animal was 1·0 and in the control animal 0·6. Similar experiments were performed by Gröber[10] and Bruns[2]. Külbs also found that the heart ratios of wild animals kept in confinement for six months were less than the heart ratios of ordinary wild rabbits.

There is a large amount of evidence regarding the influence of athletics on the heart size as estimated by the X-rays.

Numerous German and Austrian writers have published figures to show that long-continued athletics produce a well-marked increase in the size of the heart. This increase is particularly noticeable in long-distance runners, swimmers and cyclists (Diethlen[4]). This conclusion is however denied by various American writers[7].

THE INFLUENCE OF CLIMATE ON HEART RATIO

Animals living in cold climates appear to have higher heart ratios than similar animals living in warmer climates.

Strohl[27] found that the heart ratio of the alpine ptarmigan *Lagopus alpinus* was 1·63, whereas the heart ratio of the ordinary ptarmigan (*Lagopus albus*) was only 1·108.

Hesse[12] found that the heart ratio of the common sparrow (*Passer domesticus*) from St Petersburg was 1·55, whereas the heart ratio of German sparrows was only 1·384.

Hesse also examined the heart ratios of a large number of squirrels (*Sciurus vulgaris*) from various parts of Germany, and found a clear correlation between the heart ratio and the mean annual temperature of their habitat.

Mean temperature of habitat	9·3	8·3	7·7	6·6	6·3	5·9
Heart ratio	0·5	0·57	0·55	0·59	0·62	0·65

This correlation may be due to the fact that the metabolic rate of an animal in a cold climate will probably be higher than that of an animal in a warmer climate, and therefore the former animal will need to maintain the higher minute volume of circulation.

THE HEART RATIOS OF SMALL ANIMALS

Mammals and Birds below about 20 grm. in weight are all distinguished by having unusually high heart ratios.

This is particularly marked in the case of the smallest Mammals. Many of these smallest animals are however bats, and all warm-blooded animals that fly have high heart ratios.

High heart ratios occur however in other small Mammals besides bats. For example the average heart ratio of five species of mice of weights between 10 and 50 grm., measured by Hesse was 0·64, whereas the average heart ratio of four larger species of rodents of weights between 50 and 200 grm., measured by the same author, was 0·45. The relation between the body weight and heart weight in mice, rats and rabbits is shown in Fig. 12.

The heart ratios of bats as a class are higher than those of other Mammals as is seen from the following figures given by Mangan [19].

	Average body wt. in grm.	Average heart ratio
Seven species of small bats	7	1·3
Three species of larger bats	65	0·83

Similarly Strohl [27] found that in 9 specimens of *Vespertilio minimus* with an average body weight of 20 grm. the average heart ratio was 1·1.

It will be seen that the heart ratios of these bats are about twice the heart ratios of the rodents of similar size, which are shown in Table 19.

THE HEART RATIO IN LARGE BIRDS
AND MAMMALS

There are few figures for Birds of more than five kilos in weight, but Fig. 10 shows that their heart ratio is lower than that of smaller Birds, and the figures appear to follow the curve based on the formula heart weight varies as (body weight)$^{0.9}$. The ostrich is an exception, for it has a heart ratio of 1·08, but this animal is an exceptionally powerful runner.

It has been shown that when animals of the same or allied species are compared the usual relation between body weight and heart weight is:—heart weight varies as (body weight)$^{0·8}$, but that when Mammals of all sizes are compared the relationship approaches more nearly to:—heart weight varies as (body weight)$^{0·9}$.

The significance of these figures is shown by the following calculation. If the heart ratio of an animal of 10 grm. is taken as 1·0, which is slightly higher than any observed for Mammals that do not fly, then the two formulae give the following heart ratios with increasing body weight.

Table 24

Body weight in grams	(1) Heart ratio if heart wt. $\propto$ (body wt.)$^{0·8}$	(2) Heart ratio if heart wt. $\propto$ (body wt.)$^{0·9}$
10	1·0	1·0
100	0·63	0·8
1,000	0·4	0·6
10,000	0·25	0·5
100,000	0·16	0·4
1,000,000	0·1	0·3
10,000,000	0·09	0·25
100,000,000	0·06	0·2

The lowest heart ratio observed for any warm-blooded animal is that of the rabbit which is 0·27. Formula 1 reaches this limit at a body weight of less than 10 kilos. This formula therefore cannot hold over any very extensive range of body weight. Formula 2 reaches the same limit when the body weight is about 10,000 kilos. The evidence regarding the heart weight of animals of over 1000 kilos body weight unfortunately is scanty, but it indicates that the formula:—heart weight varies as (body weight)$^{0·9}$, does not hold in these cases.

Evans[6] gives the average weight of Burmese elephants as 2700 kilos, and the average weight of three hearts as 10·9 kilos. This gives a heart ratio of 0·4. Retzer[22] also mentions that the heart of a small female elephant weighed 10·5 kilos. Gilchrist (quoted from Evans) states however that the weight of the heart of an elephant which weighed 2000 kilos was 19·0 kilos. This gives

c

the surprisingly high heart ratio of 0·95 and suggests that the animal was abnormal. The heart ratio of the elephant appears to be over rather than under 0·4.

The evidence I have been able to collect regarding the hearts of whales is as follows.

Scoresby [23b] dissected a baby Greenland whale. Its length was 5·8 metres and he estimated its body weight as 5 tons (5000 kilos). The heart weighed 29 kilos, and measured 73 cm. (breadth) by 30 cm. (length) by 23 cm. (depth). He also found that the heart of a narwhal [23c] which had a body length of 4·5 metres weighed 4·9 kilos. From the body measurements given in this case I estimate the body weight to have been about one ton.

White and Kerr [28] described the heart of a whale; the animal was stated to be about 3·6 metres long; the heart measured 33 × 56 cm. and weighed 22 kilos.

The usual figures for the relation between weight and length of whales would give an impossibly high heart ratio in this case, and probably the body length was underestimated.

Murie [21] examined a rorqual 18·4 metres long, whose body weight he estimated as 45 tons; the heart measured 80 cm. base-apex, breadth at base 105 cm. and at apex 87·5 cm. I calculate that the probable heart weight was about 200 kilos.

Bouvier [1] described a specimen of *Hyperoödon rostratus*, the body measured 7·20 × 1·55 metres and the heart 57 × 47 cm. Hunter (1787) [13] stated that in a sperm whale of 60 ft. long the heart filled a large tub and he estimated the output per beat of the heart of the whale at 10–15 gallons (45–68 litres); this would be a possible figure for a heart of about 200 kilos.

This evidence is very scanty but the measurements given suggest that the heart ratio of large whales is at least 0·4 and that the heart ratio of the smaller cetacea is about 0·5.

The evidence available suggests therefore that in Mammals of more than 1000 kilos weight the heart does not decrease with increase of body weight according to the general formula found for smaller Mammals.

THE RELATION BETWEEN HEART WEIGHT
AND BODY WEIGHT

The evidence considered may be summarised as follows:

(1) Birds on an average have hearts nearly twice as large as Mammals of the same size. Amongst Mammals the bats have exceptionally large hearts, and their hearts are as large as those of birds of a similar size.

(2) In both Birds and Mammals the general average of heart ratios shows that the heart weight tends to vary as $(\text{body weight})^{0.9}$.

Very great individual variations occur, and the above rule is only a rough generalisation.

(3) In animals of the same species there is a clear relation between body weight and the heart ratio. The smaller animals have the higher heart ratios, and this is true whether adults of varying size are compared, or whether young animals are compared with adults. In many cases the heart weight varies as $(\text{body weight})^{0.8}$.

(4) Active Mammals capable of severe and continued muscular exertion have higher heart ratios than inactive animals. The highest heart ratios in Birds are found in those species which are distinguished for powers of flight or running or for loudness of singing.

(5) There is a considerable amount of evidence to show that prolonged severe exercise can cause an increase in heart size, but much of this evidence is a matter of controversy.

THE PROPORTIONS OF THE HEART

The general shape of the heart remains very similar over a wide range of sizes in Mammals, and, as is shown in Fig. 6, the heart of a rat magnified 25 diameters appears very similar to the heart of an ox.

The weight of the heart in grams is approximately equal to the cube of the length in centimetres of the cavity of the left ventricle, as is shown by the following measurements made by the author.

The relation between the weights of the right and left ventricle

6-2

Table 25

Animal		Heart weight in grm.	$\sqrt[3]{\text{heart weight}}$	Length of cavity of left ventricle
Mouse ...	...	0·13	0·47	0·55
Rat ...	...	0·67	0·87	1·0
Rabbit ...	...	5·8	1·8	2·2
Dog ...	...	102	4·65	4·0
Sheep ...	...	210	5·9	6·5
Ox ...	...	2030	12·6	12·0
Horse ...	...	3900	15·7	16·0

does not appear to alter with size. The author made a series of measurements in which the ventricles were separated from the auricles, and the right wall of the right ventricle was cut off and its weight compared with the weight of the rest of the ventricles. The following figures were obtained:

Table 26

Animal	Ratio: $\dfrac{\text{Right wall of right ventricle}}{\text{remainder of ventricles}}$
Mouse	0·3
Rat	0·2
Rabbit	0·24
Dog	0·38
Sheep and pig ...	0·3
Ox	0·325

Sainsbury [23a] who used a similar method obtained an average ratio in man of 0·4.

These figures show that there is no obvious relation between body weight and thickness of the wall of the right ventricle, but the dog which has a powerful heart has a relatively large right ventricle and the rabbit and rat which have feeble hearts have relatively small right ventricles.

Animals with powerful hearts have relatively large right ventricles, as is shown by the following figures:

Table 27

Animal	Heart ratio	Ratio: $\dfrac{\text{Right ventricle}}{\text{left ventricle}}$
Rabbit, tame [8]	0·24	0·49
,, wild [8]	0·28	0·61
Hare [8]	0·77	0·67
Duck, tame [9]	0·63	0·41
,, wild [9]	1·06	0·55
Ptarmigan		
Lagopus lagopus [27]	1·11	0·347
Lagopus alpinus [27]	1·63	0·562

The figures in Table 27 were obtained by the method devised by Müller(20) in which the interventricular septum is divided. With this method Müller obtained a ratio of 0·52 in man.

In the turkey Lewis(17) obtained the ratio R. V./L. V. = 0·3 as opposed to a ratio of 0·55 which he found in the dog.

These figures suggest that when the minute volume of the circulation is increased more extra work is thrown upon the right than upon the left ventricle: the reason for this is discussed in Chapter XIII.

The importance of this distinction is however somewhat doubtful because in man the most varying forms of heart hypertrophy occur in disease. Lewis(16) found that the most usual form of heart hypertrophy in disease was a uniform hypertrophy of both chambers. In diseases with high blood pressures however the hypertrophy affected the left ventricle particularly. In mitral disease with a normal blood pressure the right side of the heart was chiefly affected.

References

(1) Bouvier. *Ann. des Sc. Nat.* 7th Ser. Zool. B, **13**, 288. 1892.
(2) Bruns. *Münch. Med. Woch.* **56**, 1008. 1909.
(3) Buchanan. *Journ. of Physiol.* **47**; *Proc.* iv. 1913.
(4) Diethlen. *Moderne Methoden der Kreislaufs Diagnostik.* Thieme. Leipzig. 1925.
(5) Donaldson. *The Rat.* Philadelphia. 1924.
(6) Evans. *Elephants and their Diseases.* Rangoon. 1910.
(7) Gordon et al. *Arch. of Int. Med.* **33**, 425. 1924.
(8) Gröber. *Deut. Arch. f. klin. Med.* **91**, 502. 1907.
(9) Gröber. *Pflüger's Arch.* **125**, 507. 1908.
(10) Gröber. *Arch. f. exp. Path. u. Pharm.* **59**, 424. 1918.
(11) Heinroth. *Journ. f. Ornithol.* **70**, 172. 1922.
(12) Hesse. *Zool. Jahrb. Abt. Allg. Zool.* **38**, 243. 1921.
(13) Hunter, John. *Phil. Trans. Roy. Soc.* 1787.
(14) Joseph. *Journ. of Exp. Med.* **10**, 521. 1908.
(15) Külbs. *Verh. d. Kongr. f. inn. Med.* **26**, 197. 1909.
(16) Lewis. *Heart*, **35**, 367. 1913.
(17) Lewis. *Phil. Trans. Roy. Soc.* B, **207**, 221. 1916.
(18) Loer. *Pflüger's Arch.* **140**, 293. 1911.
(19) Mangan. *Compt. Rend. Soc. de Biol.* **23**, 657. 1917.
(20) Müller. *Die Massenverhaltn. d. Mensch. Herzen.* Voss, Hamburg, 1883.

(21) Murie. *Proc. Zool. Soc.* **33**, 206. 1865.

(22) Retzer. *Anat. Record*, **6**, 75. 1912.

(23) Robinson, Bryan. 1734. Quoted from Müller (20).

(23 a) Sainsbury. *The Heart as a Power Chamber.* Oxford Med. Publ. 1922.

(23 b) Scoresby. *Journ. of a Voyage to the Northern Whale Fishery*, p. 148. Edinburgh, 1823.

(23 c) Scoresby. *An Account of the Arctic Regions.* Vol. 1, p. 495. Edinburgh, 1820.

(24) Skavlem. *Amer. Journ. of Physiol.* **61**, 501. 1922.

(25) Starling and co-workers. *Journ. of Physiol.* 1912–1915.

(26) Stewart. *Amer. Journ. of Physiol.* **57**, 27. 1921.

(27) Strohl. *Zool. Jahrb. Abt. Allg. Zool. u. Physiol.* **30**, 1. 1910.

(28) White and Kerr. *Heart*, **6**, 205. 1916.

Chapter XI

THE RELATION BETWEEN CIRCULATION VOLUME AND BODY WEIGHT IN WARM-BLOODED ANIMALS

The Metabolic Rate of Animals—Variations in Pulse Frequency—Circulation Volume—Athletic and non-Athletic Animals—Small Animals—Large Animals—Animals of same Species—Influence of Age—Calculation of Pulse Rate—Adjustment between Metabolic and Pulse Rates—Birds and Mammals.

THE METABOLIC RATE OF ANIMALS

The chief and most urgent function of the circulation is to furnish oxygen to the tissues, and therefore it is necessary to consider first the manner in which the oxygen requirements are affected by variations in size.

Rubner[27] showed that in dogs of varying sizes the metabolic exchange varied as (body weight)$^{\frac{2}{3}}$; this has been confirmed recently by many workers for a very large number of Birds and Mammals and for dogs in particular by Lusk and Du Bois[17]. The surface of the body also is proportional to (body weight)$^{\frac{2}{3}}$, and it has been found that as a general rule the metabolic exchange of warm-blooded animals per unit of surface is nearly constant, being from 800 to 1000 calories per square metre body surface per day.

If the total metabolism of animals varies as (body weight)$^{\frac{2}{3}}$, then the metabolism per unit weight or metabolic rate, which is $\dfrac{\text{total metabolism}}{\text{body weight}}$ will vary as $\dfrac{(\text{body weight})^{\frac{2}{3}}}{\text{body weight}}$. That is to say the metabolic rate will be inversely proportional to $\sqrt[3]{\text{body weight}}$.

Figs. 13 and 14 illustrate the general truth of this rule, for they show that in Birds and Mammals the metabolic rates lie along a line drawn to the formula

$$\log \text{metabolic rate} = \log K - \tfrac{1}{3} \log \text{body weight.} \quad (K = 1000.)$$

The smallest Birds form an exception and their metabolic rates are from two or three times as great as the figures given by this rule.

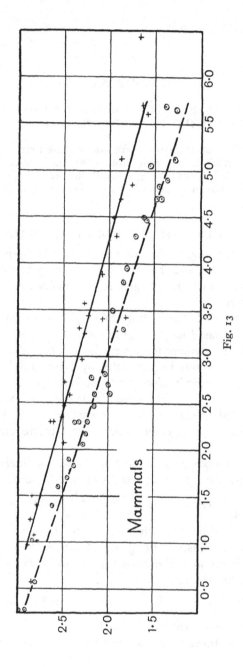

Fig. 13

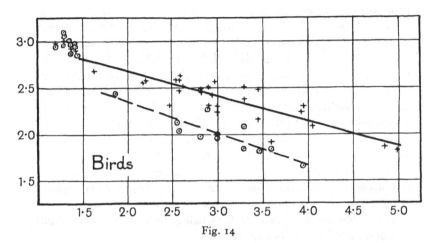

Fig. 14

Figs. 13 and 14. Influence of body weight on pulse frequency and metabolic rate in birds and mammals.

Ordinate: log of pulse frequency and of metabolic rate.

Abscissa: log of body weight in grams.

$$\frac{+}{+ \quad +} = \text{log of pulse frequency per minute.}$$

$$\frac{\odot \qquad \odot}{\odot} = \text{log of metabolic rate (calories per kilo body weight}$$

per 24 hours).

The curves are plotted from the figures shown in Appendix I.

The lines through the figures for the pulse rates follow the formula:

$$\text{Pulse rate} = \frac{K}{(\text{body weight})^{0.27}};$$

or log pulse rate $= \log K - 0.27 \log$ body weight. ($K = 1400$.)

The lines through the figures for the metabolic rates follow the formula:

$$\text{Metabolic rate} = \frac{K}{(\text{body weight})^{0.33}};$$

or log metabolic rate $= \log K - 0.33 \log$ body weight. ($K = 1000$.)

The exact significance of this rule is a matter of dispute. It has been assumed that the variations in the metabolic rates of warm-blooded animals are related to the fact that their heat loss must be proportional to their surface.

Some workers however consider that the relation between metabolic exchange and body surface is probably accidental. Rubner[28] gives figures to show that in cold-blooded animals the metabolic exchange also is proportional to the body surface and in this case the surface heat loss can play little part. Krogh[15] however concludes that there is no certain relation between metabolic exchange and heat loss in cold-blooded animals.

Benedict and Talbot[6] also doubt whether in the case of infants the heat loss is proportional to the body surface.

Recent experiments upon isolated tissues by Meyerhof and Himwich[20] prove that the variations in oxygen consumption observed in intact animals are also seen in isolated tissues.

These observers worked with small strips of the diaphragm and obtained the results shown below.

Table 28

Animal	Average body weight in grm.	$\dfrac{1}{\sqrt[3]{\text{body weight}}}$	Ratios of figures in preceding column	Oxygen consumption in c.c. per grm. per hour in bicarbonate Ringer at $38°$ C.
Large rats	220	0·1655	1·00	1·00
Young rats	72	0·241	1·46	1·45
Mice	22	0·358	2·19	2·22

These figures show that the oxygen consumption per unit weight of the isolated muscle varies inversely as the cube root of the body weight. This is the same relation as that found for the metabolic exchange of the intact animal.

Whatever the ultimate cause may be for the variations in metabolic exchange in animals of different sizes, these differences are due to the activity of the tissues being "set" at different rates and are still shown when the tissues are deprived of all nervous control.

THE FACTORS DETERMINING VARIATIONS IN PULSE FREQUENCY

A study of the isolated tissues which show rhythmic activity proves that the frequency of these tissues, which are deprived of all nervous control, varies in a manner very similar to the variations observed in the metabolic exchange. A few figures are shown in Table 29; unfortunately the activity of an organ such as the heart is interfered with considerably by isolation and therefore these figures are not accurate enough to permit of mathematical analysis. They prove however that the variations in the frequency of excised organs are of the same order as the variations of frequency *in situ* (cf. Fig. 14).

Table 29

Animal	Body weight	Frequency of heart		Frequency of intestine		Metabolic rate (calories per kilo per 24 hours)
		Isolated	*In situ*	Isolated	*In situ*	
Rat	150	350 (9)	420	35 (9) 42 (13)	—	227
Guinea-pig	250	300 (9)	300	—	—	153
Rabbit	2,000	200 (9)	205	21 (1)	21 (1)	78
Cat	2,500	153 (26)	120	12 (18)	12 (16)	80
Dog	10,000	133 (22, 32)	110	11 (18)	11 (4)	51·5

This relation between the body weight and the frequency of the isolated heart is seen in animals of the same species, for an analysis of the records of 84 experiments on the heart-lung preparation made by Starling and his co-workers (22, 32) (Table 30) shows that there is a well-marked correlation between the body weight and the frequency of the isolated heart in dogs of varying size.

Table 30

No. of observations ...	8	20	16	23	13	4
Weight of heart in grm.	30–40	41–50	51–60	61–70	71–100 (av. 85)	101–150
Approx. wt. of dog in kilos	4	5	6·6	8·0	11·5	18
Average frequency of heart	151	151	136	141	135	121

In this case the variation in the pulse rate is less than would be the case if the pulse rate varied as $\dfrac{1}{\sqrt[3]{\text{body weight}}}$.

The metabolic exchange and the frequency of isolated tissues vary in a very similar manner, moreover they are affected in an identical manner by changes in temperature (as is shown in Fig. 8).

It is therefore probable that there is some common factor which determines the metabolic exchange and frequency of rhythm of tissues. The most probable explanation appears to be that the frequency of rhythm depends upon the rate of metabolic exchange of the tissues.

The variations in pulse frequencies of different animals are due therefore fundamentally to the pacemaker of the hearts of different animals being set at different rates, just as the variations in metabolic exchange are due to differences in the tissues.

The pulse rate in the living animal is however under the control of the vagus and sympathetic nerves, and can be varied rapidly to meet the fluctuating needs of the body in the same way that the heat exchange also can be varied by nervous influences.

Since the pulse rate of an animal is affected by every emotion and every activity it is difficult to obtain figures for pulse rates which admit of accurate comparison.

The basal pulse rate of an animal is the rate in a fasting animal, at complete bodily rest and under no emotional stress. Such measurements are only obtainable from tame animals.

The pulse rates of a number of Birds and Mammals are shown in Figs. 13 and 14 in which the body weights, pulse rates and metabolic rates are plotted logarithmically. The observations on which these tables are based are contained in Appendix I.

Figs. 13 and 14 show that the variation in pulse rate is not parallel to the metabolic rate, and that the larger animals have more rapid pulse rates than would be anticipated from the differences in metabolic rates. This is to be anticipated from the fact that the heart weight is not proportional to the body weight but to (body weight)$^{0.9}$.

The oxygen supplied to the tissues by the circulation depends on three factors: (i) the amount of oxygen supplied by a unit volume of blood, (ii) the output per beat of the heart, and (iii) the frequency of the heart beat.

There is no evidence that any extensive variation occurs in the first factor in animals at rest and the other two factors alone need consideration. The output of the heart is presumably proportional to its weight and this varies as (body weight)$^{0\cdot9}$. Hence the output per beat of the heart per unit of weight varies inversely as (body weight)$^{0\cdot1}$.

The metabolic rate varies inversely as (body weight)$^{0\cdot33}$, since the metabolic rate indicates the oxygen used per unit weight of tissue, and the oxygen supplied to the tissues depends on the heart frequency and heart output per beat per unit weight of tissues. Therefore the pulse rate × heart output per unit of weight should vary as the metabolic rate.

Therefore pulse rate $\propto \dfrac{\text{metabolic rate}}{\text{heart output per unit weight}}$.

Therefore pulse rate $\propto \dfrac{(\text{body weight})^{0\cdot1}}{(\text{body weight})^{0\cdot33}}$.

Hence log pulse rate
$$= \log K + 0\cdot1 \log \text{body weight} - 0\cdot33 \log \text{body weight}$$
$$= \log K - 0\cdot23 \log \text{body weight}.$$

The lines through the pulse rates in Figs. 13 and 14 follow the formula log pulse rate $= \log K - 0\cdot27 \log$ body weight. ($K = 1400$.)

The observed pulse rates fit therefore approximately to the formula and the deviation between the calculated and observed variation is probably due to errors caused by the great individual variations in the heart ratios.

These calculations show that the variations in the amounts of oxygen per unit of time and weight required by animals of varying sizes are met chiefly by variations in the pulse rate.

The fact that the uncontrolled frequency of the pacemaker of the heart in animals of different sizes varies with the oxygen consumption per unit weight of the heart provides a mechanism which ensures a correlation between the circulation volume per minute and the oxygen needs of animals.

THE CIRCULATION VOLUME

The correlation between pulse frequency and metabolic rate is only approximate, and therefore it is necessary to consider the various factors involved in the provision of oxygen to the tissues in more detail.

The circulation in resting animals only has been hitherto considered. It is obviously advantageous to an animal to provide a circulation during bodily rest as economically as possible, but it is much more important for most animals to have an apparatus which will provide efficiently for emergencies. The survival of most animals depends more on their power to meet emergencies than upon minor economies effected during periods of tranquillity.

The maximum circulation which the heart is capable of providing appears to be the chief limiting factor in the performance of sustained muscular exertion.

The oxygen supply as already mentioned depends on three circulatory factors, namely oxygen utilisation, systolic output and frequency. All these factors are of importance in providing an increased oxygen supply during exertion. Their relative importance is indicated by the following figures obtained by Yandell Henderson [14] from a trained athlete.

Table 31

	Standing at rest	Violent exertion (1500 kg.m.) work per min.	Ratio of increase
1. Litres oxygen used per minute ...	0·597	4·6	7·7
2. Oxygen utilisation coefficient (difference between c.c. of O_2 per 100 c.c. of arterial and of venous blood)	6·1	15·16	2·5
3. Pulse rate	68	152	2·2
4. Stroke volume in litres (calculated from 1, 2 and 3)	0·144	0·197	1·36

This table shows that all the three circulatory factors mentioned are of importance in providing increased oxygen supply during exertion. When animals of different size at rest are compared the pulse rate is the factor which varies in much the most obvious manner. The smallest Birds and Mammals have pulse rates approaching from 600 to 1000, whilst the largest have pulse rates from 35 to 70. The stroke volume probably is proportional to the heart weight, and this is influenced by size, since on the average it varies as (body weight)$^{0.9}$, but it depends chiefly on the mode of life of the animal; in Mammals the heart ratio varies from 0·27 (rabbit) to 1·6 (small bats), and in Birds from 0·5 to 1·8.

As regards oxygen utilisation sufficient figures are not available to permit any decision as to whether the proportion of available oxygen used varies greatly in different animals at rest. No relation has been demonstrated between the haemoglobin content of the blood and body weight, but the figures for active animals such as the horse, dog, cat and man show about 15 per cent. of haemoglobin in the blood, whereas the figures for non-athletic animals such as the rabbit, ox, goat and sheep vary from 10 to 12 per cent. Yandell Henderson found that dogs and men at rest used 15 to 20 per cent. of the oxygen content of the blood, and Barcroft[2] found that goats used 25 per cent.

ATHLETIC AND NON-ATHLETIC ANIMALS

The term athletic animals is used to indicate animals which need to perform severe and continuous muscular exertion.

The hare and rabbit form an extremely good example of the contrast between the circulation in these two classes of animals.

Table 32

	Rabbit	Hare
Heart ratio	0·27	0·77
Pulse rate at rest	205	60–70
Pulse rate after vagal section ...	321	264

These figures show the contrast at once; the hare has a heart three times the size of that of a rabbit the same size, and at rest the hare's pulse rate is only one-third that of the rabbit. The slow pulse rate of the hare is however due to vagal control, and when this control is removed the heart rates are similar. Simple abolition of vagal control therefore enables the hare to increase its circulation per minute 4-fold, whereas the rabbit by this means can only produce an increase of 1·6-fold. Athletic animals as a class have large hearts, slow pulses and a low oxygen utilisation coefficient when at rest. This is an arrangement which permits a maximum increase in oxygen supply to the tissues during violent exertion.

In men during violent exertion the oxygen supply to the tissues can be increased to 16-fold the resting value, whereas a rabbit probably can only produce about a 3-fold increase.

It is interesting to note that the low pulse rates of athletic animals are not due to the fundamental frequency of the heart's pacemaker being low, but are due to the powerful vagal control.

The fact that athletic animals have large hearts and consequently slow pulse rates explains most of the pulse rates that fall far away from the averages in the case of Mammals and Birds of medium size.

The pulse rates of the smallest Mammals and Birds follow the general rule as regards the relation between body weight and pulse rate. In the case of the smallest Birds this is remarkable because their metabolic rate is two to three times as great as would be anticipated from the general rule relating metabolic rate to body weight. These small Birds all have remarkably high heart ratios, and thus provide the extra circulation needed. This suggests that a frequency of about 1000 a minute is the highest at which the vertebrate heart can function efficiently. A probable explanation for this limit is that it is difficult to see how the coronary circulation could supply sufficient oxygen to the heart to maintain a greater rate.

Information regarding the pulse rates of the largest animals is very scanty. The pulse rate of the elephant (3000 kilos) is the same as that of the ox (500 kilos), namely 45–50, and greater than that of the horse (500 kilos), which is 35–40. This suggests that there is a lower limit to the pulse rate, below which the circulation does not work efficiently.

THE CIRCULATION VOLUME OF ANIMALS OF THE SAME SPECIES BUT OF VARYING BODY WEIGHTS

The figures given in Table 20 show that in several animals of the same species the weight of the heart varies as (body weight)$^{0.8}$.

The increased metabolic needs of small animals of the same species are therefore very largely met by variation in the heart ratio and a consequent variation in the stroke volume per kilo body weight. For example, if dogs of widely varying size are compared the following results are seen:

Table 33

	Small dog (5 kilos)	Large dog (30 kilos)	Ratio
Calories per kilo per 24 hours	76	39·7	1·93
Heart ratio (indicating stroke volume per kilo)	0·88	0·56	1·57

These figures indicate that not much difference in pulse rate in dogs of different size is to be expected, since the ratio of pulse rates needed to completely compensate the different metabolic rates is only $\dfrac{1\cdot93}{1\cdot57} = 1\cdot23$, i.e. if the large dog in Table 33 had a pulse rate of 100 the small dog would have a pulse rate of 123.

Unfortunately, no reliable figures are available showing the differences in pulse rates of adult animals of the same species but of different sizes, but such figures are available for animals during growth.

The metabolic rates of new-born animals are much higher than those of adult animals, and this difference is compensated for partly by a higher heart ratio and partly by a higher pulse rate.

Table 34

	Infant 1 year old 10 kilos wt.	Adult 20 years old 70 kilos wt.	Ratio
Metabolic rate, cals. per kilo per 24 hours	56 (6)	24	2·34
Pulse rate	116 (6)	64	1·8
Heart ratio	0·6	0·52	1·16

In this case it will be seen that there is no very extensive difference between the heart ratios, and the higher circulatory needs of the infant are met principally by the higher pulse rate. The amount of variation in the heart ratio differs in different species. Fig. 12 shows that in the young rat and in infants of less than 10 kilos the heart weight ratio varies approximately as (body weight)$^{0\cdot8}$.

The pulse rates and metabolic rates of various new-born animals are shown in Table 35.

Table 35

The Heart Rates of Young Animals

	Weight in grm.		Heart rate		Metabolic rate	
	New-born	Adult	New-born	Adult	New-born	Adult
Mice	1·15 (24)	25	—	600	780 (24)	600
Guinea-pigs	64·5 (24)	500	—	200	206 (24)	54
Cats	117 (24)	2,000	300 (9)	120	189 (24)	80
Dogs	401 (24)	14,000	180 (19)	100	108 (24)	60
Men	3,000	70,000	112 (6)	64	60 (6)	24
Horses	—	400,000	100 (11)	30	—	—
Oxen	—	500,000	160 (11)	45	—	—

Accurate figures are available for the pulse rate and metabolic rate of human infants, and it is interesting to note that the corelation between the variation in the pulse rate and metabolic rate is only approximate. In the first few months after birth the metabolic rate rises considerably, whilst there is no corresponding rise in the pulse rate, and the heart ratio falls slightly. The reason for this is unknown but the divergence would be explained if newborn infants had a low coefficient of oxygen utilisation.

CALCULATION OF PULSE RATE

Buchanan[8 a] pointed out that the pulse rates of mammals and birds could be calculated approximately from the metabolic rate and heart ratio. This relation can be expressed

$$\text{Pulse rate} = K \frac{\text{metabolic rate}}{\text{heart ratio}},$$

or, since the metabolic rate varies as $(\text{body weight})^{0.66}$,

$$\text{Pulse rate} = \frac{K}{\text{heart ratio} \times (\text{body weight})^{0.33}}.$$

Table 36 shows a series of calculations made according to this formula. $(K = 1400.)$

In the case of Birds, Figs. 10 and 11 show that the average heart ratios of Birds are 1·4 times as great as Mammals of the same size, but Figs. 13 and 14 show that the average pulse rates and metabolic rates of Birds and Mammals of the same size are approximately the same.

In the case of Birds therefore the formula used to calculate pulse rates was

$$\text{Pulse rate} = \frac{K \times 1·4}{\text{heart ratio} \times (\text{body weight})^{0.33}}.$$

The agreement between the figures given by the formula and the observed pulse rates is only approximate. This is to be expected as several variables have been left out of consideration, for example, the body temperature of animals varies considerably and, as Figs. 13 and 14 show, the basal metabolism of individual animals varies considerably from the standard curve.

In the Mammals, the calculated figures for the rabbit and hare are twice the observed figures; this suggests that in these animals

Table 36

Animal	Body weight in grm.	Heart ratio	Pulse rate calculated	Pulse rate observed
Mammals				
Vesperugo pipistrellus	4	1·4	620	660
Mouse ...	25	0·65	710	670
Rat ...	200	0·48	490	420
Guinea-pig ...	300	0·45	466	300
Rabbit ...	2,000	0·27	405	205
Hare	3,000	0·77	126	64
Dog	5,000	0·88	94	120
,,	30,000	0·56	80	85
Sheep	50,000	0·45	85	70
Man	70,000	0·52	65	72
Horse	450,000	0·64	29	38
Ox	500,000	0·47	37	40
Elephant ...	3,000,000	0·45	22	48
Sperm whale ...	100,000,000	0·4	88	—
Birds				
Canary ...	20	1·5	470	1000
Pigeon ...	300	1·5	196	185
Crow	341	0·9	313	378
Buzzard ...	658	0·77	293	300
Hen	2,000	0·55	278	312
Domestic duck	2,300	0·63	232	240
Wild duck ...	1,100	1·06	180	190
Turkey ...	8,750	0·8	118	193
Ostrich ...	71,000	1·08	43	60–70

the oxygen utilisation from the blood may be exceptionally high. The general rule gives far too low a figure for elephants, and the figure for the largest whales seems very improbable. This confirms the suggestion already made that none of the general laws regarding metabolic rate, pulse rate and heart weight hold for Mammals over about 1000 kilos in weight. In Birds the agreement between calculated and observed figures is fairly good except in the smallest Birds; in these the metabolic rates are twice as great as would be anticipated from the general rule relating to body weight and metabolism, and similarly the observed pulse rate is twice as great as the calculated figure.

Considering that the formula has been applied over more than a million-fold range of weight, the agreement between calculated and observed figures must be considered as fairly satisfactory.

THE ADJUSTMENT BETWEEN METABOLIC
RATE AND PULSE RATE

From a study of infants, Benedict and Talbot (5) concluded that whatever raised the pulse rate also increased the metabolism.

Extensive studies of the relation between pulse rate and metabolic rate have been made in cases of thyroid disease in which the metabolic rate varies from 40 per cent. below to 60 per cent. above normal (23, 25), and it has been found that the pulse rate varies in a manner parallel to the metabolic rate. This association is so constant in individuals of a similar size that it has been suggested that the basal metabolic rate can be estimated from the pulse rate.

Benedict and co-workers (7, 8) showed in both cattle and men that prolonged underfeeding lowered the basal metabolic rate and also caused a lowering of the pulse rate.

Table 37
Average pulse rates and metabolic rates

		Under-nutrition	Normal feeding	Fattening
Oxen (8)	Cals. per kilo per 24 hours	16	24	—
	Pulse rate	33	44·5	78
Men (7)	Cals. per kilo per 24 hours	23·1	25·2	—
	Pulse rate	42	55–56	—

The chief mechanisms for varying the pulse rate to meet an emergency such as muscular exercise or emotional stress are the removal of vagal control and the excitation of the sympathetic cardiac centre. This mechanism is very rapid, for the pulse rate increases within a few seconds of the onset of muscular exercise or excitement.

The facts mentioned above suggest however that there is some delicate reflex mechanism that adjusts the degree of vagal control in the resting animal so that the pulse rate is sufficient to maintain an adequate circulation volume and an adequate oxygen supply to the tissues.

In man the change of posture from lying to standing somewhat hinders the venous return of blood to the heart and reduces the stroke volume, and this change at once causes an increase in the pulse rate of 10 to 15 beats a minute.

A further proof of the fact that the pulse rate at rest is adjusted to supply an adequate oxygen supply to the brain is provided by observations on carbon monoxide poisoning or reduced oxygen tension.

Low oxygen tension causes a temporary reduction in the metabolic rate which returns to normal in a few hours (29). It causes a well-marked increase in the pulse rate however (10, 12, 29). Carbon monoxide poisoning also causes an increase in the pulse rate (21).

After prolonged exposure to lowered atmospheric tension the pulse rate returns to normal at rest but is accelerated enormously by any slight exertion (3, 31).

The evidence regarding the relation between oxygen tension in the tissues and pulse rate is therefore inconclusive, but it suggests that some relation does exist.

THE RELATIVE EFFICIENCY OF THE HEARTS OF BIRDS AND MAMMALS

Figs. 13 and 14 show that the metabolic rates of Birds and Mammals of equal weight are approximately equal, and so also are their pulse rates.

Figs. 10 and 11 show however that the heart ratios of Birds are about 1·4 times as great as those of Mammals of the same size.

This suggests that a Bird requires 1·4 times as much heart muscle as a Mammal to supply the tissues with a given amount of oxygen, and that the Bird's heart has to do more work than a Mammal's heart to produce the same effect.

This difference is apparently due to two factors: firstly, the Bird's blood is a less perfect oxygen carrier than the Mammal's, and, secondly, the Bird requires a higher blood pressure than the Mammal to drive the blood through its tissues. The Bird's blood is similar to that of non-athletic Mammals, for it contains about 10 per cent. of haemoglobin (30), and the arterial blood contains 10–15 volumes per cent. of oxygen (33). The blood in athletic Mammals contains however 14 per cent. haemoglobin and the arterial blood contains 20 volumes per cent. of oxygen. In their general mode of life Birds must be compared to athletic Mammals for most Birds are capable of very prolonged and severe exercise.

The blood pressure in Birds is about one and a half times that of Mammals of similar size. These differences appear sufficient to account for the relatively greater heart weight in Birds.

References

(1) Alvarez. *Amer. Journ. of Physiol.* 31, 267. 1915.
(2) Barcroft et al. *Quart. Journ. of Med.* 13, 35. 1920.
(3) Barcroft et al. *Phil. Trans. Roy. Soc.* B, 211, 351. 1923.
(4) Bayliss and Starling. *Journ. of Physiol.* 24, 99. 1899.
(5) Benedict and Talbot. *Publ. of Carnegie Instit. of Washington*, 201. 1914.
(6) Benedict and Talbot. *Publ. of Carnegie Instit. of Washington*, 233. 1915.
(7) Benedict et al. *Publ. of Carnegie Instit. of Washington*, 280. 1919.
(8) Benedict and Ritzman. *Publ. of Carnegie Instit. of Washington*, 324. 1923.
(8 a) Buchanan. *Science Progress*, 5, 60. 1910.
(9) Clark. Unpublished observations.
(10) Doi. *Journ. of Physiol.* 55, 43. 1921.
(11) Ellinger, 1894. Quoted from Tigerstedt. *Physiol. d. Kreisl.* 2, 468. 1921.
(12) Hambach and Linhard. *Bioch. Zeit.* 68, 265. 1915, and 74, 1. 1916.
(13) Hansmell. *Amer. Journ. of Physiol.* 40, 48. 1922.
(14) Henderson, Yandell. *Amer. Journ. of Physiol.* 73, 193. 1925.
(15) Krogh. *The Respiratory Exchange of Animals.* London, 1916.
(16) Langley and Magnus. *Journ. of Physiol.* 33, 34. 1903.
(17) Lusk and Du Bois. *Journ. of Physiol.* 59, 213. 1924.
(18) Magnus. *Pflüger's Arch.* 102, 123. 1904.
(19) Meyer. *Arch. de Physiol.* 5, 475. 1893.
(20) Meyerhof and Himwich. *Pflüger's Arch.* 205, 415. 1924.
(21) Mosso. *Arch. Ital. de Biol.* 35, 21. 1901.
(22) Patterson and Starling. *Journ. of Physiol.* 48, 357 and 465. 1914.
(23) Peterson and Walker. *Journ. Amer. Med. Ass.* 78, 341. 1922.
(24) Plant. Quoted from Winterstein's *Handb. d. vergl. Physiol.* 2, ii. 1037. 1924.
(25) Read. *Arch. of Int. Med.* 34, 553. 1924.
(26) Rohde and Ogawa. *Arch. f. exp. Path. u. Pharm.* 69, 200. 1912.
(27) Rubner. *Zeit. f. Biol.* 19, 535. 1883.
(28) Rubner. *Bioch. Zeit.* 148, 222. 1924.
(29) Schneider et al. *Amer. Journ. of Physiol.* 70, 283. 1924.
(30) Schultz and Krüger. Winterstein's *Handb. d. vergl. Physiol.* 1, i. 1118. 1924.
(31) Somervell. *Proc. Roy. Soc. Med. Sec.* 11. 1922.
(32) Starling and Evans. *Journ. of Physiol.* 46, 67. 1914.
(33) Winterstein. *Handb. d. vergl. Physiol.* 1, ii. 1. 1924.

Chapter XII

HEART FUNCTIONS IN COLD-BLOODED ANIMALS AND EMBRYOS

Metabolism and Pulse Frequency—Fish Hearts—Amphibia and Reptiles
Blood Pressures—Circulation in Invertebrates—Embryos.

METABOLISM AND PULSE FREQUENCY

The relation between metabolic rate, pulse rate and body weight is very clear in warm-blooded animals, but the relationship is doubtful in cold-blooded animals.

In the first place there is no certain relation between the body weight and metabolic rate of cold-blooded Vertebrates.

Krogh(15), in summarising the conflicting evidence on this question, concluded, "In reptiles, amphibia and fishes, values of 0·2–0·5 calories per kilo per hour are generally observed on quiet animals irrespective of size." He pointed out that small animals are more active than large animals and hence the metabolic rates found for young, small and active animals are usually higher than those found for larger and quieter animals.

On the other hand, Rubner(21) concluded that the rule which states that the metabolic exchange per unit area of surface is constant was true for Fishes as well as for warm-blooded animals; he gave the following figures for Fishes:

Table 38

	Body weight in grm.	Cals. per kilo per diem at 16° C.
Stickleback ...	1·75	39
,,	2·75	31
Goldfish ...	4	12
Tench ...	250	4·7
Salmon ...	9,500	2·95
Sturgeon ...	1,400,000	0·26 (estimated)

Rubner also gave the following figures for metabolic exchange per unit of surface in cold-blooded Vertebrates.

Table 39

	Body weight in grm.	Cals. per sq. metre per diem	
Fish	1·0	40	
,,	4·0	26	· Average figure 33
,,	200	32	
Frogs	1·3–600	97·5	
Lacerta	110	45	
Tortoise	135	64	
Alligator	1380	47	
Mammals and Birds	All sizes above 50 grm.	1000	

Table 40

Cold-blooded Vertebrates

Animal	Body weight in grm.	Heart ratio $\frac{heart\ wt. \times 100}{body\ weight}$	Pulse rate at about 16° C.	Metabolic rate ° C. cals. per kilo per hour
Cyclostomes				
Petromyzon	298	0·277 (9)	—	—
Elasmobranchs	—	0·075–0·12 (9)	—	—
Scyllium catullus	1000	—	44	—
Carcharias (shark)	100 cm. long	—	30 (15a)	—
	200 cm. long (circa 40 kilos)	—	18 (15a)	—
Teleosteans	—	0·03–0·12 (9)	—	—
Stickleback	2	—	60–100	30 (21)
Goldfish (Carrasius)	4	—	—	12 (21)
Eel	36 cm. long	—	40–60 (14)	—
Cod (Gadus morrhua)	30 cm. long	—	26–40 (23)	—
Pike (Esox lucius)	500	—	30–42 (12)	—
Amphibia				
Rana temporaria	25	0·278 (9)	30	13 (21)
,, mugiens	600	—	—	125 (21)
Reptiles				
Lacerta viridis	14	0·220 (9)	60–66 (5)	25 (21)
Tortoise	135	—	10–20 (7)	8·5 (21)
Crocodile	71	—	22–47 (5)	—
Tropidonotus natrix	169	0·320 (9)	23–41 (5)	—

FISH ·HEARTS

The frequency of the hearts of fish embryos is relatively low. The following figures are given by St Paton (18) for the frequency of *Pristiurus* embryos of varying lengths: 5 mm., 16–20; 10 mm., 27–29; 17 mm., 43–45.

Preyer (19) gives the frequency of freshly-hatched trout as 50–72 and St Paton that of trout embryos of 15 mm. as 75–80, but Babuk and Hepner(1) give the frequency of the heart in freshly-hatched trout as about 100 at 15° C.

The figures available for the frequencies of fish hearts are so scanty and irregular that it is difficult to generalise upon them. There seems however to be no clear relation between pulse frequency and body weight such as would be anticipated if Rubner's rule held good that the metabolic rate was proportional to the body surface. If this rule were true and the pulse frequency varied as the metabolic rate, then a Fish of 1 grm. should have a pulse frequency ten times as great as a Fish weighing 1000 grm., but no such variations are observed.

With regard to the heart ratio, this is much lower than that of Mammals. This is explicable by the fact that the oxygen requirements of a Fish are only about one-thirtieth those of a warm-blooded animal of the same size, and hence a much smaller circulation volume per minute must be adequate for a Fish.

AMPHIBIA AND REPTILES

The information regarding the pulse rates of these animals also is scanty. The pulse rate of snakes was found by Buchanan to vary very greatly and to increase rapidly as soon as the animal became restless.

The metabolic rate of these animals is from one-tenth to one-twentieth that of warm-blooded animals of the same size. The haemoglobin content of frog's blood is similar to that of Mammals' (10 per cent.). The heart ratio of Amphibia and Reptiles is about one-half that of the Mammals. These figures show that a pulse frequency of less than a tenth that of a Mammal of the same size should be sufficient to provide the oxygen needs of the Amphibia and Reptiles. In actual fact the pulse frequencies observed are between one-tenth and one-twentieth of those of Mammals of the same size.

THE BLOOD PRESSURE OF COLD-BLOODED VERTEBRATES

The blood pressure of cold-blooded Vertebrates is lower than that of warm-blooded animals. Table 41 shows a number of

collected figures for the blood pressure of cold-blooded Vertebrates.

It will be seen that in very few cases does the blood pressure
rise above 50 mm. Hg.

In the cases of Fishes the blood pressure in the branchial
arteries is given.

Table 41

Blood pressures of cold-blooded Vertebrates [11]

Animal	Blood pressure in mm. Hg	
Carcharias (shark)	32	Lyon [15a]
Torpedo	16–18	Schönlein
Scyllium	30–37	,,
Raja	20	Hyde
Esox lucius	35–84	Volkmann
Salmo	47–120	Greene
Barbus fluviatilis	42	Volkmann
Anguilla	65–70	Legerot and Jolyat
Coluber natrix	89	Hofmeister
Bufo terrestris	48	,,
Rana esculenta	c. 40	Schultze, Kuno
,, temporaria	37–55	Klug, Jacobi, Straub
,, mugiens	5–30	Hofmeister
Tortoise (Pseudemys rugosa)	18–35	Edwards
,, (Testudo)	30–50	Legerot and Jolyat
Crocodile	30–50	Hofmeister

THE CIRCULATION IN INVERTEBRATES

The Invertebrates resemble the cold-blooded Vertebrates in
that there is no clear relationship between size and pulse frequency.

There has been much controversy regarding the relation between
body weight and metabolic rate in Invertebrates, and many authors
have found that the total metabolism varies as (weight)$^{0.66}$.
Montuori [16], however, made a very extensive series of observations on a large number of cold-blooded animals and found that
the metabolism per unit weight was constant, irrespective of the
size of the animal. Krogh [15] concluded that the metabolic exchange per unit weight was more or less constant in Invertebrates,
but that very young animals often showed much higher values
than normal, probably to a great extent on account of their greater
muscular activity. The hearts of Invertebrates differ too much in
structure to permit of many comparisons being drawn. As a rule
hearts of small animals have a higher frequency than those of large
animals. For example, the pulse rate of small Crustacea weighing

a few mgm. is 100–200 whilst that of lobsters and crabs (500–1000 grm.) is 50–100. Similarly, the frequency of the heart of a fly is much greater than that of a cockroach.

Schultze [22] found that in Tunicates the frequency of the hearts of large species was 26–30, whilst that of small species was 107.

HEART FUNCTIONS IN EMBRYOS

The embryonic heart in Birds and Mammals supplies the tissues with oxygen, but the oxygen needs of the tissues in embryos are very different from those of free-living forms, since the temperature of the foetus is maintained by maternal warmth.

Hence the exact relationships found in warm-blooded animals between body surface, metabolic rate and pulse rate are not to be expected in the embryo or foetus.

The Size of the Embryonic Heart

Figures for the relation between the heart weight and body weight are available in the case of the human foetus.

Heart Ratio of human foetus (Müller [17]).

	Foetus			Infant		
	1–2	2–3	3–4	1–2	2–3	3–4
Weight in kilos ...						
Heart ratio	0·622	0·645	0·619	0·65	0·60	0·615

These figures show that the heart ratio reaches a maximum about birth and afterwards diminishes. The most remarkable point is how slightly the figures vary during foetal and extra-uterine life, since the circulation in the foetus is so different from that in the adult.

The Pulse Rate of Bird Embryos

The following rates [20] have been observed for the hen's embryo during incubation:

Days' incubation	2	3	4	6	8	11
Pulse rate ...	90	122	123	128	148	167

These figures agree with those of various other observers and it appears that the pulse rate of the embryo increases steadily during incubation. Unfortunately no figures are available for the pulse rates of newly-hatched chickens, but the pulse rate of the adult hen is about 320, which is twice that of the embryo. This is inter-esting because the metabolic rate of the embryo chick is far greater than that of the adult hen.

Bohr and Hasselbach (4) found for example in an egg at the 10th day of incubation a metabolic rate of about 100 calories per kilo per 24 hours as compared with a figure of 60 for the adult hen. Their figures for younger embryos are higher still. The embryo in the egg therefore has a higher metabolic rate and a lower pulse rate than the adult animal.

The Embryonic Rate in Mammals

The following figures show that the rate of the foetal heart is slightly greater than that of the new-born animal:

Table 42

	Foetus	New-born	Adult
Human ...	135 (5–9 months) (10)	112 (2)	70
Oxen ...	161 (6)	141 (6)	50
Dog ...	120–170 (8)	160 (8)	100

This variation in pulse rate does not correspond with variations in metabolic rate, for in the case of the guinea-pig the following metabolic rates (calories per kilo per 24 hours) have been found: Foetus 58·5 (3); New-born 206 (13); Adult 54.

Birth apparently causes a rapid increase in the metabolic rate of Mammals, a fact also noted by Hill in the case of rats. Neverthe-less, there does not appear to be any corresponding increase in the pulse rate.

References

(1) Babuk and Hepner. *Folia Neuro-biologica*, 6, 368. 1912.
(2) Benedict and Talbot. "Physiology of New-born Infants." *Publ. of Carnegie Instit. of Washington*, **233**. 1915.
(3) Bohr (1900). Quoted from Krogh, *The Respiratory Exchange of Animals*. London, 1916.
(4) Bohr and Hasselbach. *Skand. Arch. f. Physiol.* 10, 149. 1900.
(5) Buchanan. *Journ. of Physiol.* 39, Proc. xxv. 1909.
(6) Ellinger (1884). Quoted from Tigerstedt, *Physiol. d. Kreisl.* 2, 468. 1921.
(7) Gaskell. *Journ. of Physiol.* 4, 43. 1883.
(8) Henriques. *Zeit. f. Biol.* 26, 197. 1887.

(9) Hesse. *Zool. Jahrb.*, Abt. All. Zool. **38**, 243. 1921.
(10) John and Schick. *Zeit. f. Kinderheilk.* **38**, 216. 1924.
(11) Junk. *Tabulae Biologicae.* **1**, 143. 1925.
(12) Kazem-Beck and Dogiel. *Zeit. f. wiss. Zool.* **37**, 247. 1882.
(13) Kestner and Plant. Winterstein's *Hand. d. vergl. Physiol.* **2**, ii. 1037.
       1924.
(14) Kolff. *Pflüger's Arch.* **122**, 37. 1908.
(15) Krogh. *The Respiratory Exchange of Animals.* London, 1916.
(15 a) Lyon. *Journ. of Gen. Phys.* **8**, 279. 1926.
(16) Montuori. *Arch. Ital. de Biol.* **59**, 213. 1913.
(17) Müller. *Massenverh. d. Mensch. Herz.* Voss, Hamburg. 1883.
(18) Paton. *Mitt. a. d. Zool. Stat. in Neapel*, **18**, 544. 1909.
(19) Preyer (1885). Quoted from Tigerstedt, *Physiol. d. Kreisl.* **2**, 74.
       1921.
(20) Preyer and Kolliker. Quoted from Tigerstedt, *Physiol. d. Kreisl.* **2**,
       475. 1921.
(21) Rubner. *Bioch. Zeit.* **148**, 222, 268. 1924.
(22) Schultze. *Jena. Zeit. f. Naturw.* **35**, 232. 1901.
(23) Thesen. *Arch. de Zool. exp. et gén.* **3**, 3, 122. 1896.

Chapter XIII

THE WORK OF THE HEART IN RELATION TO BODY WEIGHT IN WARM-BLOODED ANIMALS

The work of the Heart—Blood Pressure—Velocity of Blood—Aortic Cross-section—Duration of Systole—Calculation of Velocity of Blood—Work of the Heart—Coronary Flow.

THE WORK OF THE HEART

The hearts of all warm-blooded animals are of very similar proportions and it is of interest to consider what is the influence of alteration of dimensions on the manner in which the heart performs its work.

The heart performs mechanical work in two ways, first by expelling at each beat a certain volume of blood against arterial resistance and secondly by imparting to this ejected blood a certain velocity.

The total rate of work is the sum of the two and is given by the expression (Evans [7])

$$W = QR + \frac{MV^2}{2g} \text{ for each ventricle.}$$

Let Q = vol. of blood in c.c. expelled per second,

R = blood pressure in terms of cm. of blood,

M = mass of blood ejected in grams per second,

V = velocity at which it is ejected in cm./sec.,

g = gravity constant (981 cm. per sec.[2]).

Then W = gram centimetres of work performed in unit time.

This calculation shows that the work of the heart may be divided into the static factor QR and the kinetic factor $\frac{MV^2}{2g}$.

The relative importance of these two factors depends upon the velocity of the blood stream, since the static factor varies as the output of the heart per minute and therefore directly as the velocity of the blood leaving the heart, while the kinetic factor varies as (velocity)^3. Hence, if the output of the heart is increased

10-fold, the static factor (QR) is also increased 10-fold, but the kinetic factor $\left(\dfrac{MV^2}{2g}\right)$ is increased 10×10^2, that is, 1000-fold.

When calculating the velocity it is not sufficient to estimate the mean aortic velocity, but the velocity of blood during systolic output must be calculated.

The velocity of blood during systolic output can be calculated if we know the minute output of the heart, the cross-section of the aorta and the proportion of the cardiac cycle occupied by systolic output.

Any calculation of the work of the heart should measure the work done by the whole heart. The work done by the auricles is however negligible compared with the work done by the ventricles. In chapter x it was shown that the relation between the mass of the right and left ventricles was approximately the same in animals of widely different sizes, and therefore the work done by the left ventricle alone will be considered, and it will be assumed that the relation between the amounts of work done by each of the two ventricles is similar in animals of varying sizes. The weight of the left ventricle will be taken as one-half of the total weight of the heart.

The following figures give an example of how the work done by the heart can be calculated.

Yandell Henderson [13] found in an athlete that the circulation in litres per minute was 9·8 at rest and 30·3 during violent exertion.

The cross-section of the aorta (84-kilo individual) is about 5 sq. cm., and the duration of systole at rest and at work may be taken as 30 and 40 per cent. of the cardiac cycle respectively. These figures give an aortic velocity during systolic output of 109 and 252 cm. per second respectively.

If the blood pressure be taken as equal to 120 mm. Hg = 155 cm. blood, the following figures for the work of the heart per minute are obtained:

(1) *At rest*

$$W = 9800 \times 155 + \frac{9800 \times 1 \cdot 055 \times (109)^2}{1962}$$
$$= 1{,}520{,}000 + 66{,}000.$$

(2) *During exercise*

$$W = 30{,}300 \times 155 + \frac{30{,}300 \times 1 \cdot 055 \times (252)^2}{1962}$$

$$= 4{,}700{,}000 + 1{,}040{,}000.$$

This shows that in a man's heart the kinetic factor represents less than one-twentieth of the total work done while at rest, but during violent exercise it amounts to one-fifth of the total work done.

It will be shown later that in small animals the velocity of the blood is greater than in large animals, and therefore the kinetic factor is relatively greater in small animals.

This varying relation between the static and kinetic factors of the work done is of interest for the comparison of the work done by the two sides of the heart.

The volume output of the two sides of the heart is the same, and since the cross-sections of the pulmonary artery and aorta are almost the same, the velocity of the blood in the two arteries must be the same. The pressure in the pulmonary artery is however only about one-sixth that in the aorta. The static factor of work will therefore be six times as great on the left as on the right side of the heart, but the kinetic factor will be equal on the two sides.

In large animals at rest therefore the left ventricle will do six times as much work as the right ventricle, but in small animals at rest, or in large animals during violent exercise, the difference will be much less. In the calculations made above for the work of the heart during violent exercise in man, the work done by the right side of the heart must have been about one-third of that done by the left side. This fact at once explains why violent exercise causes hypertrophy of the right side of the heart.

In order to calculate and compare the work done by the heart in animals of different sizes it is necessary to know the following factors:

(1) The volume of blood expelled from the left ventricle per minute.

(2) The systolic blood pressure.

(3) The specific gravity of the blood.

(4) The velocity at which the blood is ejected.

All these factors can be either measured or calculated.

THE CIRCULATION VOLUME PER MINUTE

This can be calculated from the oxygen consumption of the animal provided that the amount of oxygen yielded to the tissues by a unit volume of blood is known. Unfortunately this factor has only been measured in a few animals.

As was shown in Table 31 the oxygen utilisation coefficient (the amount of oxygen provided to the tissues by a unit volume of blood) is a variable factor, that can be increased about two and a half times during violent exercise.

When animals of different species are considered there are two possible variables, namely, the total amount of oxygen per 100 c.c. blood and the proportion used by the tissues.

The total oxygen content of the blood does not however appear to vary very greatly, for the haemoglobin content of healthy warm-blooded animals appears to be remarkably constant. In athletic Mammals it is about 15 per cent. and in non-athletic Mammals and in Birds it is about 10 to 12 per cent. As regards the proportion of oxygen used, Yandell Henderson[13] found that in man and the dog at rest 4 c.c. oxygen were utilised per 100 c.c. arterial blood and Barcroft[2] obtained a figure of 5 c.c. in the case of the goat. Yandell Henderson considers that as a general average an athletic Mammal at rest utilises about 15–20 per cent. of the oxygen of the arterial blood and that in non-athletic Mammals this figure is somewhat higher. Hence it appears that at rest the amount of oxygen supplied by the arterial blood to the tissues may be taken as about 5 c.c. oxygen per 100 c.c. blood.

SYSTOLIC BLOOD PRESSURE

The blood pressure attained during systole does not vary very greatly in animals of varying sizes. It is higher in large animals than small ones and is higher in Birds than in Mammals.

A few typical figures are given in Table 43.

Volkmann concluded that the blood pressure did not vary with the body weight of the animal. This is not strictly true, but the body weight has remarkably little influence, since a horse which is 2000 times as heavy as a rat has a blood pressure less than twice as high. The reason why the variation is so slight is probably that

c

Table 43

	Weight in grm.	Artery	Blood pressure. Height in mm. Hg
Birds			
Jackdaw ...	140	Carotid	119 (21)
Turkey ...	9,000	,,	193 (21)
Mammals			
Bat ...	circa 20	Wing	50 (14)
Rat ...	200	Carotid	100 (Clark)
			71–80 (19)
Guinea-pig	300	,,	81–90 (19)
Dog ...	10,000	,,	100–140 Av. 120 (15)
Horse ...	400,000	,,	155 (25)

the resistance in the capillaries is similar in all Mammals. The specific gravity of the blood does not appear to vary greatly in warm-blooded animals of different size, and may be taken on the average as 1·055.

THE VELOCITY OF BLOOD IN THE AORTA

A certain number of direct measurements of the velocity of the blood flow in the large arteries have been made. The following figures are quoted from Tigerstedt [24]:

Table 44

		Velocity in cm. per sec.			
Animal	Artery	max.	min.	Method of measurement	Author
Horse	Carotid	43	22	Haemodromometer	Volkmann
	,,	52 (syst.)	15 (diast.)	,,	{ Chaveau, Bertolus, Laroyenne
Calf	,,	43	—	,,	Volkmann
Sheep	,,	35	24	,,	,,
Goat	,,	36	24	,,	,,
Dog	,,	36	20	,,	,,
	,,	38	10	,,	Tschuewsky
	,,	34	10	Haemotachometer	Vierordt
Rabbit	,,	34	10	Haemodromometer	Jensen
	Aorta	26 (average)		,,	Tigerstedt

These figures indicate that during systole a velocity of from 35–40 cm. per sec. is attained and that this velocity is similar in animals varying in size from the horse to the rabbit. Unfortunately these direct measurements are of somewhat doubtful value because their recording involves considerable operative interference.

The velocity of blood during systole also can be calculated if we know the circulation volume per second, the cross-section of the aorta and the percentage of the cardiac cycle that is occupied by systoles. Unfortunately the figures for the circulation volume are somewhat uncertain because they have to be calculated from the oxygen consumption and the oxygen utilisation, and this last factor is only known for certainty in the case of man, dogs and goats.

THE AORTIC CROSS-SECTION

Dreyer Ray and Walker[6] measured the aortic cross-section in rats and guinea-pigs of varying sizes and found that in animals of a single species the aortic cross-section varied as (body weight)$^{0.71}$. In the rat Donaldson's[4] figures show that the heart weight varies

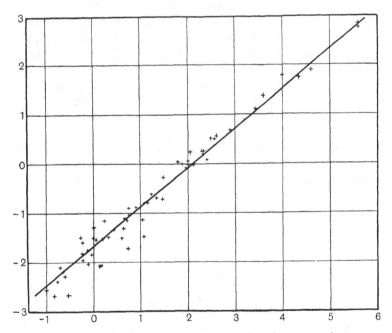

Fig. 15. Relation between heart weight and aortic cross-section.
    Ordinate: log. of aortic cross-section in square cm.
    Abscissa: log. of heart weight in grams.
    The curve is drawn following the relation:
    Aortic cross-section varies as (heart weight)$^{0.8}$.

approximately as (body weight)$^{0.8}$, and therefore the aortic cross-section appears to vary in this case as the (heart weight)$^{0.9}$.

The writer has collected and made measurements of the aortae of a large number of animals. The results are shown in Fig. 15, and it will be seen that over a 10,000,000-fold range of body weight the aortic cross-section varies as (heart weight)$^{0.8}$. The heart weight, on the average, varies as (body weight)$^{0.9}$ and therefore the aortic cross-section varies approximately as (body weight)$^{0.72}$.

DURATION OF SYSTOLIC OUTPUT

In order to find the average velocity during systole the time of systolic output must be known. The proportion of heart cycle occupied by systole in any species depends on the frequency of the pulse.

For example, Roos[20] found in one case in man that during rest with a pulse rate of 69 the systole occupied 29 per cent. of the cardiac cycle, but that during an attack of tachycardia the pulse rose to 160 and the time occupied by systole rose to 43 per cent.

Yandell Henderson[12] found similarly in dogs that when the heart rate was varied from 90 to 200 per minute the duration of systole varied very little, and hence the greater the frequency the greater the proportion of the cardiac cycle that was occupied by systole.

The following figures show the proportion of the cardiac cycle occupied by systolic output in various animals: most of them have been calculated by the writer from records of aortic and ventricular pressures reproduced in Tigerstedt[22]:

Table 45

Animal	Pulse rate per minute	Percentage of cardiac cycle occupied by systolic output	Author
Horse	40	27	Chaveau and Marie
Man	46	23	Roos
	160	43	Roos
	80	30	Roos
Dog	50	20	Hürthle
	85	35	Hürthle
	130	40	Piper
Cat	162	43	Piper
Rabbit	240	40	Tigerstedt
Rat	320	40	Clark (calculated from electrocardiograph)

These figures show that the proportion of the cardiac cycle occupied by systole is greater in small animals with rapid pulse rates than in large animals with slow pulse rates.

In man Lompard and Cope [17] calculated that the duration of systole varied inversely as $\sqrt[2]{\text{pulse rate}}$.

CALCULATION OF VELOCITY OF BLOOD IN AORTA

The figures obtained permit the calculation of the velocity which the blood must attain in the aorta during systole.

Table 46

Calculation of aortic velocity

1. Animal	2. Body weight in grm.	3. Calories per kilo per 24 hours	4. Minute volume of blood in c.c.— calculated from col. 3	5. Duration of systole as percentage of cardiac cycle	6. Aortic cross-section in sq. cm.	7. Velocity of blood in aorta during systole in cm. per sec.— calculated from cols. 4, 5 and 6
Mouse	25	400	29	40	0·004	300
Rat	200	210	122	40	0·02	255
Rabbit	2,000	76	440	40	0·1	184
Dog	10,000	50	1,450	35	1·05	66
Sheep	50,000	27·4	3,980	(35)	1·76	108
Man	70,000	25	5,070	30	4·0	71
Horse	500,000	20	29,000	27	23·7	76
Ox	500,000	24	34,800	(27)	12·6	170

The figures in column 4 were estimated on the assumption that 100 c.c. of blood provide 5 c.c. of oxygen to the tissues, and that 1 litre of oxygen yields 4·8 calories.

They can be checked by the calculation of Yandell Henderson [13] that in man and the dog the output of the heart is about 1·5 c.c. per kilo per beat when at rest, this would give for a dog of 10 kilos, with a pulse rate of 100, a minute volume of 1500 c.c., and for a man of 70 kilos, with a pulse rate of 65, a minute volume of 6370 c.c. Other observers [5, 16] have obtained for a 70 kilo man at rest minute volumes varying from 4700 to 7500 c.c. Most of the figures shown in Table 46, column 7, for the velocity of blood in the aorta are about twice as great as those found by direct measurement and shown in Table 44. This is probably due to the fact that the direct measurements were made in the carotid. In the case of the rabbit the calculated value is no less than six times the observed value, but obviously the interference with circulation involved in introducing a "strom-uhr" into the aorta must render the circulation very abnormal.

It has been shown that the oxygen requirements of animals vary as (body weight)$^{0.66}$, and the cross-section of the aorta varies as (body weight)$^{0.71}$. As body weight increases therefore the cross-section of the aorta increases slightly more rapidly than does the minute volume of blood needed.

All the figures for velocities in Table 46 have been calculated on the assumption that 5 volumes of oxygen are used from every 100 c.c. blood. Probably this figure is too high in the case of athletic animals such as the horse, man and dog and therefore the aortic velocity in these animals is somewhat greater than calculated.

On the other hand, the calculated aortic velocity in the mouse and rat seems improbably high, and it is possible that in these animals the oxygen utilisation coefficient is higher than 5 volumes per cent. The figures as a whole indicate that the velocity of the blood in the aorta during systolic output of the heart and during bodily rest is about 100 cm. per second.

During exercise, however, this velocity must be far greater. Yandell Henderson [12] has shown that when the heart is accelerated the duration of systole does not shorten greatly and hence the proportion of time occupied by systole rises. This reduces the velocity somewhat, but in the extreme case in man of the oxygen consumption being raised 16-fold above the resting value by violent exercise, the circulation volume must be increased at least 6-fold and the velocity in the aorta during systolic output at least 4-fold.

During violent exercise, therefore, at the time when the heart is doing its maximal work the aortic velocity must approach 400 cm. per second. The figures for circulation in man during exercise, quoted on page 111, show an aortic velocity during systolic output of 252 cm. per second.

THE WORK OF THE HEART IN ANIMALS OF VARYING SIZES

The figures available permit the calculation of the work done per gram of muscle of left ventricle per minute in hearts of varying sizes. It was shown in chapter x that the proportion between the weights of the right and left ventricles was similar in hearts of all sizes, and therefore the weight of the left ventricle is taken as being 0·5 times the weight of the heart.

The work of the hearts of various animals at bodily rest calculated according to the formula given on page 110 and the figures in Table 46 give the following results:

Table 47

Animal		Work in grm.cm. per minute per grm. of left ventricle
Mouse	... ...	50,000
Rat	... ...	45,000
Rabbit	... ...	21,000
Dog	... ...	6,000
Man	... ...	4,000
Horse	... ...	5,600

The figures in Table 47 can be checked by direct measurements made on the isolated heart-lung preparation of the dog. Evans and Matsuoka [7] found that a dog's heart weighing 46·5 grm. did 180 kg.m. work per hour when the output was 81 litres per hour and the arterial pressure was 80 mm. Hg. A dog with a heart of this size would require a cardiac output nearly as great as this to maintain its normal metabolic rate, and the work done corresponds to 6500 grm.cm. work per minute per gram of heart. On the assumption that the left ventricle did five-sixths of the work this would give a figure of 10,800 grm.cm. per minute per gram of left ventricle. Figures given by Tigerstedt [23] correspond to about 12,000 grm.cm. per minute per grm. of left ventricle in the rabbit.

The chief interest of these figures is that they show that the heart of a mouse at bodily rest must do more work per gram of substance than does the heart of a large Mammal during violent exercise. This raises the interesting problem as to how the hearts of small Mammals can obtain enough oxygen to carry out their violent and sustained activity.

THE CORONARY FLOW

The only reliable information concerning the coronary flow is that derived from experiments on the heart-lung preparation of the dog.

Anrep(1) has shown that the coronary flow in dogs, when the nervous supply to the heart is intact, is chiefly dependent on the blood pressure. Any increase in blood pressure produces a very great increase in coronary flow. The output of the heart affects the coronary flow to a lesser extent, and the rate of the heart has little effect on the coronary flow. Evans and Starling(8) showed that the coronary flow in the dog's heart-lung preparation was about 60 per cent. of the heart weight per minute. Anrep's figures show that in conditions approximating to a normal resting animal the coronary flow is about 4 per cent. of the total output. The coronary flow is, however, doubled by a rise of blood pressure such as is produced by adrenaline.

Evans and Starling(8) found that the coronary flow in the dog's heart-lung preparation under the influence of acidosis might amount to 20 per cent. of the total output of the heart, and attain the figure of 370 c.c. per minute per 100 grams heart weight.

The coronary flow depends in the first place on the aortic pressure, but in addition the coronary vessels are provided with vaso-constrictor nerves from the vagus and vaso-dilator nerves from the sympathetic (Anrep(1)).

Evans and Ogawa(9) found that the oxygen utilisation of the coronary blood flow in the heart-lung preparation was about 6 volumes of oxygen per 100 c.c. of blood under normal conditions, and rose to 13 volumes of oxygen after adrenaline.

The coronary flow in the dog is therefore only just adequate to supply the necessary amount of oxygen to the heart.

The coronary flow in small animals must be very great. In a mouse of 25 grm. the heart appears to perform about 50,000 grm.cm. work per gram per minute. Supposing an efficiency of 30 per cent., this involves the utilisation of 0·8 c.c. oxygen per gram per minute (1 litre oxygen = 4·8 cals. and 1 cal. = 425·5 kg.m. work). Supposing the 10 volumes of oxygen are provided by 100 c.c. blood, this means a blood flow of 8 c.c. per gram. per minute, a volume only paralleled by the blood supply of the ductless glands in the larger Mammals.

These figures suggest that the mouse's heart, when the animal is at bodily rest, is working almost up to the limit of the possible oxygen supply, and indicate a reason why Mammals of less than

10 grm. weight have disproportionately large hearts. The smaller the animal the higher its oxygen requirements and therefore the larger the volume of circulation per minute per gram of body weight. Down to a certain body weight these increasing requirements are met by increasing the frequency of the heart beat, but there is a limit to the possible pulse rate because there is a limit to the amount of blood that can be supplied to the heart per minute per gram of tissue. This limit appears to be reached at a body weight of between 10 and 20 grm., and below this weight the increased blood supply needed is provided by increasing the relative size of the heart.

THE WORK OF THE HEART DURING BODILY EXERCISE

In the previous discussion chief attention has been paid to the work of the heart during bodily rest. The power of an animal to sustain a severe muscular exertion depends however chiefly on the reserve powers possessed by its heart. As has already been mentioned, animals vary greatly in the reserve powers of the heart, but in athletic animals these are very great.

The figures given on page 111 show that during violent muscular exertion in man the heart output can increase 3-fold, and this involves a $3\frac{1}{2}$-fold increase in the work done by the heart and a $2\frac{1}{2}$-fold increase in the velocity of the blood during systole.

The reserve powers of the hearts of small animals must be very much less, for their normal activity involves an oxygen consumption which approaches the limit that can be supplied by the coronary flow.

THE INFLUENCE OF BODY WEIGHT ON ARTERIAL RESISTANCE

The work done by the heart is almost entirely spent in overcoming friction in the arteries and capillaries, for the blood enters the veins at a very low pressure and velocity.

The size of the capillaries is very similar in all animals, but since the blood flow through a unit weight of tissue is far greater in small than in large animals, it is possible that the capillary resistance varies considerably in animals of different size.

The work necessary to drive the blood through the arteries must however be much greater in large than in small animals. The smaller arteries are presumably similar in all animals and therefore the work needed to drive the blood through the larger arteries in a large animal is additional work which is not required of the smaller animals.

The resistance due to friction when a fluid flows through a tube is in the first place proportional to the viscosity of the fluid and the length of the tube, and is inversely proportional to the square of the radius of the tube. The relation between resistance and velocity is a complex one. When the flow is of stream line nature the resistance varies as the velocity, but when the flow is turbulent the resistance varies more nearly as the square of the velocity.

The critical velocity at which turbulent flow commences depends on the viscosity of the liquid and the diameter of the tube.

Müller (18) found experimentally that when blood flowed through an elastic tube of 7·5 mm. diameter the flow began to be turbulent at a velocity of about 50 cm. per second. The flow in the large arteries of the larger Mammals must therefore be of a turbulent nature, and even in the dog the flow must be turbulent during violent exercise.

Any increase in output of the heart and consequent increase in velocity must therefore raise considerably the resistance to the blood flow through the arteries.

This rapidly increasing resistance involved by an increase in blood velocity in large animals owing to the occurrence of turbulent flow suggests a reason why it is important for such animals to have as low a blood velocity as possible, and is probably the reason why the cross-section of the aorta increases with increasing body weight more rapidly than does the minute volume of the blood.

The actual amount of work done in forcing the blood through the larger arteries can be estimated from measurements of blood pressure in different arteries.

Dawson (3) found in the dog that when the pressure in the anonyma was 123 mm. Hg the pressure in the smallest arteries measured was 117 mm. Hg.

In the horse and the ox however the fall of pressure is much greater. The aortic pressure in these animals is about 160 mm. Hg, but the pressure in the tail artery of the horse is only about 85 mm. Hg (Fontaine (10)), and in the ox 106 mm. Hg (Götze (11)). These figures must be corrected for difference in height, since in the horse the tail artery is about 35 cm. above the heart, which corresponds to a blood pressure of 26 mm. Hg. There must be however a true difference of 30–50 mm. Hg between the pressures in the aorta and tail artery in these animals.

The smallest Mammals have a blood pressure of about 70 mm. Hg, and it appears as if this was about the pressure needed to force the blood through the capillaries. The figures given suggest that in the horse at rest half of the work done by the left ventricle is expended in overcoming the friction in the larger arteries: this factor must be increased when the velocity is increased during violent exercise owing to the rise in the cardiac output per minute.

References

(1) Anrep. *Physiol. Rev.* 6, 596. 1926.
(2) Barcroft et al. *Quart. Journ. of Med.* 13, 35. 1920.
(3) Dawson. *Amer. Journ. of Physiol.* 15, 244. 1906.
(4) Donaldson. *The Rat.* Philadelphia, 1924.
(5) Douglas and Haldane. *Journ. of Physiol.* 56, 69. 1922.
(6) Dreyer Ray and Walker. *Journ. of Physiol.* 44; *Proc.* xiv. 1912.
(7) Evans. *Recent Advances in Physiology*, p. 119 ff. London, 1925.
(8) Evans and Starling. *Journ. of Physiol.* 46, 413. 1913.
(9) Evans and Ogawa. *Journ. of Physiol.* 47, 446. 1914.
(10) Fontaine. *Arch. f. (Anat. u.) Physiol.* 217. 1919.
(11) Götze. *Berl. Tierärztl. Wochenschr.* 36, 293. 1920.
(12) Henderson, Yandell. *Amer. Journ. of Physiol.* 16, 325. 1906.
(13) Henderson, Yandell. *Amer. Journ. of Physiol.* 73, 193. 1925.
(14) Hill, L. *Journ. of Physiol.* 54; *Proc.* cxliv. 1921.
(15) Hoskins and Wheelon. *Amer. Journ. of Physiol.* 34, 81. 1914.
(16) Krogh and Linhard. *Skand. Arch. f. Physiol.* 27, 100. 1912.
(17) Lompard and Cope. *Amer. Journ. of Physiol.* 49, 140. 1919.
(18) Müller. *Zeit. f. d. ges. exp. Med.* 39, 210. 1924.
(19) Porter and Richardson. *Amer. Journ. of Physiol.* 23, 131. 1908.
(20) Roos. *Deut. Arch. f. Klin. Med.* 92, 327. 1908.
(21) Stübel. *Pflüger's Arch.* 135, 249. 1910.
(22) Tigerstedt. *Physiol. d. Kreisl.* 1, p. 213, etc. 1921.
(23) Tigerstedt. *Physiol. d. Kreisl.* 3, 118. 1922.
(24) Tigerstedt. *Physiol. d. Kreisl.* 3, 166. 1922.
(25) Zuntz and Hagemann. Quoted from Tigerstedt. *Physiol. d. Kreisl.* 3, 153. 1922.

Chapter XIV

THE INFLUENCE OF CHEMICAL ENVIRONMENT ON THE HEART'S ACTIVITY

Effect of Changes in Ionic Concentrations on Frog's Heart—On Hearts of Birds and Mammals—On Fishes' Hearts—On Invertebrate Hearts—Mollusca—*Limulus*—Crustacea—Insecta—Electrolyte Content of Blood.

The discovery by Ringer in 1874 that an excised frog's heart could continue to function normally for a prolonged period, if perfused with a fluid containing a suitable content of sodium, potassium and calcium, was the foundation of a new branch of physiology. These original discoveries were gradually extended and general laws have been established regarding the nature of the environment necessary for the normal functions of cells. The elaboration of the technique of tissue cultures is one of the most important developments of this branch of physiology.

Isolated hearts, and particularly those of cold-blooded Vertebrates, have proved exceptionally easy material on which to study the effects produced by variations in the composition of the perfusion fluid, and many of the general laws now found to be applicable to many other varieties of cells, were discovered by experiments on the isolated heart.

It is now known that isolated vertebrate tissues can function normally when immersed in a fluid satisfying the following conditions. The fluid must be isotonic with the body fluids of the animal, and must contain the kations Na, K and Ca in the molecular proportions of about 100 : 2 : 1. Furthermore the fluid must be feebly alkaline $C_H = 10^{-7} - 10^{-8}$, and must contain a buffer substance. Finally there must be adequate oxygen tension in the fluid.

THE ACTION OF IONS ON THE FROG'S HEART

The heart of the frog or of the tortoise functions normally when perfused with Ringer's fluid of the following molar composition: NaCl 0·11, KCl 0·002, CaCl$_2$ 0·001, NaHCO$_3$ 0·001 ($C_H = 4 \times 10^{-8}$; $\Delta = -0·45°$ C.).

The sodium chloride can be replaced in part by any non-electrolyte that does not enter the cells and is physiologically inert; the heart, for example, continues to function normally if the sodium chloride percentage is reduced to 0·3 or even 0·2, provided that the osmotic pressure is maintained by the addition of cane sugar.

Potassium can be replaced completely by rubidium and imperfectly by caesium. Zwaardemaker (39) claims that the addition of any radio-active substance can partly or completely replace potassium. The completeness of this latter substitution effect is however a matter of dispute.

Calcium can be replaced almost completely by strontium, but by no other kation.

Calcium and potassium act as antagonists within certain limits, and independent variation of the concentration of either constituent produces well-marked effects, but if the concentration of both kations is varied equally in the same direction the concentration can be increased or reduced about 3-fold without producing any very obvious effects.

Greene (1898) found that turtle heart strips contracted for a long time in isotonic glucose. Cousy (6) has shown moreover that the full activity of the frog's heart can be maintained for many hours by a solution of glucose, isotonic with 0·6 per cent. NaCl, to which about 0·05 per cent. sodium bicarbonate has been added. This effect, however, is probably due to the sponge-like structure of the frog's heart and the viscosity of the sugar solution, which result in traces of frog's serum remaining for a long time in the interstices of the heart. Dennis (10) found that diffusion from a watery solution into a sugar solution was much slower than diffusion into water.

The frog's heart is very sensitive to any increase in the hydrogen-ion concentration; and any increase above $C_H = 10^{-7}$ results in rapid injury to the heart. Decrease in the hydrogen-ion concentration has relatively little effect, for the heart functions equally well in fluids with $C_H = 4 \times 10^{-8}$ and $C_H = 4 \times 10^{-9}$, but injurious effects occur when the hydrogen-ion concentration is reduced below $C_H = 10^{-9}$. The influence of the hydrogen-ion concentration on the heart is modified by the nature of the oxygen supply; for when the hydrogen-ion concentration is increased the heart is much more readily affected by decrease in the oxygen supply.

ACTION OF ANIONS

Lussana[23] showed that small quantities of sulphate, bromide or iodide ions produced no effect on the frog's heart, but that larger doses diminished the excitability. Franca[12] found that fluoride was toxic to the heart, but that Br, I, NO_3, SO_4 and NO_3 ions did not produce any toxic action. Sakai[30] investigated the effect of various ions on the isolated frog's ventricle. He used mixtures in which the sodium chloride in Ringer's solution was replaced by isotoxic solutions of varying salts. He found that it was possible to replace the whole of the Cl ions in Ringer's fluid by other ions and still to obtain regular and nearly normal contractions. He found that as regards amplitude of contractions the ions investigated formed the following series:

$$Br > NO_3 > Cl > CH_3.COO > SO_4 > \text{tartrate} > \text{citrate}.$$

As regards spontaneous frequency this was greater with I, Br, NO_3 and SO_4 than with Cl, but much less with CH_3, COO, lactate, tartrate and citrate.

The effect of changes in the anion content of Ringer's fluid on the frog's heart is therefore much less pronounced than the effects produced by changes in the kation content.

DETAILED EFFECTS OF IONS ON THE HEARTS OF THE FROG AND THE TORTOISE

Our knowledge regarding the effect of ionic changes is far more complete regarding the hearts of the frog and of the tortoise than regarding the hearts of other animals, and therefore it is convenient to summarise our knowledge on this subject as a basis for comparison when discussing other tissues. The extent of the work on the subject makes it difficult to summarise and only the outstanding facts will be considered.

The heart functions which can be measured most accurately are as follows:

(1) Spontaneous rhythmicity.
(2) Irritability as measured by electrical stimulation.
(3) Rate of conduction of the wave of excitation.
(4) Force of contraction.
(5) Duration of the process of contraction and relaxation.

(6) Extent of chemical change as measured by oxygen consumption or heat production.

(7) Length of refractory period, i.e. rate of restoration of normal activity after a contraction.

(8) Tonus of resting muscle.

All these functions are to a large extent interdependent and therefore exact analysis is very difficult. The difficulty of analysis is further increased by the fact that as regards its reaction to ionic changes the frog's heart contains at least four separate tissues whose reactions vary both quantitatively and qualitatively; these tissues are the sinus, the auricle, the tissue connecting auricles and ventricle, and the ventricle.

The difference between these tissues can be shown most clearly by a consideration of the spontaneous rhythmicity of the tissues.

The spontaneous rhythmicity is in the first place dependent on the temperature (cf. p. 64), and also on the tension in the resting muscle (cf. p. 34).

With constant temperature and pressure the spontaneous irritability of the sinus is influenced in the following manner by ionic changes. The percentage ionic concentrations at which the spontaneous frequency of the sinus is greatest are as follows:

$$\text{NaCl } 0{\cdot}6, \text{ KCl } 0{\cdot}016, \text{ CaCl}_2 \text{ } 0{\cdot}01{-}0{\cdot}003.$$

Any decrease in NaCl concentration decreases the sinus frequency; reduction of the NaCl concentration to $0{\cdot}152$ halves the frequency, and reduction below this abolishes spontaneous contractions (Waller [35]).

The following changes all produce about 10 per cent. reduction in the frequency, namely: 4-fold increase of calcium, reduction of potassium to a quarter, and increase of hydrogen-ion concentration to C_H 3×10^{-7}.

A comparison of these reactions of the sinus with the effects of changes in ionic concentration on the spontaneous rhythmicity of auricles and ventricle, shows that the sinus functions best in a fluid rich in sodium, potassium, and hydrogen ions, and that the ventricle requires a lower sodium, potassium and hydrogen-ion content and a higher calcium content.

Martin found that the concentrations of potassium required to arrest the turtle's sinus, auricle, and ventricle were in the ratio of 6 : 2 : 1. Kolm and Pick[22] showed a similar difference in the auricle and ventricle of the frog.

Sakai[31] showed that the following were optimal percentage concentrations for the spontaneous frequency of the sinus and ventricle:

		Sinus	Ventricle
NaCl	...	0·6	0·1
KCl	...	0·01	0·005
CaCl₂	...	0·0065	0·03

Dale and Thacker[8] found that the limits of automatic frequency for the chambers of the frog's heart were as follows: sinus $C_H = 10^{-4}$ to 3×10^{-10}; auricle $C_H = 3 \times 10^{-6}$ to 3×10^{-11}; ventricle $C_H = 3 \times 10^{-7}$ to 10^{-11}.

These facts are interesting, since they show that the optimal ionic concentrations for the activity of isolated tissues differ for different tissues of the heart of a single animal.

The most striking effects of changes in ionic concentrations may be summarised as follows:

(1) Reduction of sodium chloride content to one half. This change depresses the rate of conduction in the turtle's ventricle (Pond[29]), and in the a.-v. tissue of the frog (Daly and Clark[9]), but increases the force of contraction of the frog's auricle and ventricle (Daly and Clark), and increases the duration of the mechanical response.

(2) Increase in potassium content. This change has a very powerful depressant effect on the rate of conduction in the tortoise's auricle (Seliškar[33]) and also in the a.-v. connection in the frog (Daly and Clark[9]). The force of contraction of the auricle and ventricle is greatly diminished, the refractory period is prolonged.

The CO_2 production of the heart is decreased. The excitability of the turtle's ventricle is also decreased (Schultz[32]).

(3) Decrease in calcium concentration. This causes a great decrease in the force of contraction of the heart. Conduction in the auricle is but slightly affected (Seliškar) but the a.-v. conduction is depressed (Mines[26], Daly and Clark). The excitability of the turtle's ventricle is increased (Schultz).

(4) Decrease in potassium content or increase in calcium content. These two changes produce similar effects, the chief being an

increase in the duration of the mechanical response (Daly and Clark). Excess of calcium is said to decrease the duration of the refractory period, but the author found that either decrease of potassium or increase of calcium content slightly increased the refractory period.

The effects produced by these changes vary greatly with the condition of the heart; little improvement is produced in the activity of a fresh heart, but in an exhausted heart all its functions may be improved by either reduction in potassium or increase in calcium.

Calcium and potassium are within limits antagonists, and the effects that changes in their ionic concentrations produce on the mechanical response of the heart appear very similar (e.g. lack of calcium and excess of potassium both produce a similar depressant action). Measurements of the electrical variation show, however, that potassium excess has a marked effect on the rate of conduction and profoundly alters the electric response of the heart, whereas lack of calcium has relatively little effect on conduction, and a heart in which all movement visible to the unaided eye has ceased may still produce apparently normal electrical variations.

The potassium ion appears therefore to influence especially the conduction of the wave of excitation, whereas the calcium ion has a particularly strong influence on the mechanical response.

(5) Changes in hydrogen-ion concentration. Increase in hydrogen-ion concentration depresses the rate of conduction and the force of contraction, and prolongs the refractory period.

The effect is produced only slowly, however. An alteration in the potassium or calcium content will produce half its maximum action in a few seconds, whereas a moderate change in the hydrogen-ion concentration takes nearly a minute to produce half its maximum action.

This difference suggests that there is a profound difference between the mode of action of potassium and of the hydrogen ion, although an increase of either produces a somewhat similar end effect.

THE NEED OF THE HEART FOR ORGANIC SUBSTANCES

The frog's heart will continue to beat for several days when isolated and perfused with Ringer's fluid. Its activity is somewhat

c   9 improved by the presence of 0·1 per cent. glucose which it is apparently able to metabolise.

The heart is not, however, in equilibrium with its surroundings, and all of its activities if measured will be found to be declining steadily with the passage of time.

The heart, after prolonged perfusion, passes into a hypodynamic state in which it can be stimulated by many substances which have little or no effect on the fresh heart. The lipoids of blood serum have particularly powerful stimulant action on the hypodynamic frog's heart and similar effects can be produced by crude lecithin or even by soaps of the higher fatty acids.

Increase in the calcium content moreover has a much stronger stimulant action on the exhausted than on the fresh heart. The writer has suggested that the activity of the heart is dependent on the presence on the surface of its cells of a calcium-lipoid complex which is gradually washed away by perfusion (5).

THE EFFECT OF IONIC CHANGES ON THE HEARTS OF BIRDS AND MAMMALS

These hearts have the same ionic requirements as those of Amphibia and Reptiles, and the chief difference is that a higher osmotic pressure is needed ($\Delta = -0.56° - 0.6°$ C.). The usual Locke's fluid for perfusion of warm-blooded tissues has the following molar composition: NaCl 0·156, KCl 0·0057, $CaCl_2$ 0·002, $NaHCO_3$ 0·002. About half the NaCl can be replaced by cane sugar and therefore this fraction only serves to maintain the necessary osmotic pressure.

The composition of Locke's fluid is similar to that of the serum of Mammals and Birds. This has the following molar ionic content: Mammals—NaCl 0·138–0·145, KCl 0·005–0·0062, $CaCl_2$ 0·002–0·0027. The composition of Birds' serum is the same except that it contains three times as much potassium salts.

THE ACTION OF IONS ON FISHES' HEARTS

As regards the saline content of their serum, the Teleosts form a striking contrast to the Elasmobranchs.

The Teleosts resemble the higher Vertebrates in that the saline content of their serum is fixed and is but little influenced by the saline content of their medium. The blood of the Elasmobranchs, on the other hand, is of the same osmotic pressure as the medium in which they live. Their blood, however, is not of the same saline composition as the sea, but contains a much lower content of sodium chloride, and the osmotic pressure of the blood is kept level with that of the sea water by the presence of large quantities of urea. This contrast between Elasmobranchs and the higher Vertebrates is shown in the following table of figures taken from Winterstein's handbook (37, 38).

Table 48

Osmotic pressure of blood (Δ)

	Marine (sea water $\Delta = -2\cdot1^\circ$ C)	Land and fresh water
Mammals ...	$-0\cdot65$ to $-0\cdot83$	$-0\cdot55$ to $-0\cdot69$
Birds	$-0\cdot65$ to $-0\cdot69$	$-0\cdot55$ to $-0\cdot66$
Reptiles ...	$-0\cdot60$ to $-0\cdot66$	$-0\cdot44$ to $-0\cdot60$
Amphibia ...	—	$-0\cdot40$ to $-0\cdot56$
Teleosts ...	$-0\cdot61$ to $-1\cdot04$	$-0\cdot49$ to $-0\cdot64$
Ganoids ...	—	$-0\cdot52$ to $-0\cdot76$
Elasmobranchs	$-2\cdot1$ to $-2\cdot4$	—

This contrast is further shown by the fact that the elasmobranch blood contains $1\cdot8$ per cent. of urea, as compared with the content of about $0\cdot02$ per cent. found in the Teleosts and higher Vertebrates.

This different blood composition is reflected in the behaviour of the isolated hearts.

The isolated heart of the Teleost functions well when perfused with mammalian Ringer (Beresin(2)). It is interesting to note that the Cyclostomes appear to resemble the Teleosts in the composition of their blood, since their hearts when isolated function with frog's Ringer.

The elasmobranch heart, however, requires a totally different medium. Baglioni and other writers have devised formulae, and the writer found a fluid of the following molar composition satisfactory: Urea $0\cdot33$, NaCl $0\cdot4$, KCl $0\cdot006$, $CaCl_2$ $0\cdot012$, P_H $8\cdot3$.

The essential points are that the fluid must have the same osmotic pressure as sea water, but that about one-third of this pressure must be maintained with urea or some other inert non-electrolyte such as cane sugar. The ionic composition of the perfusion fluid differs from that of sea water in that the former contains no magnesium.

The following is the approximate ionic composition of sea water expressed as molar concentration: NaCl 0·6, KCl 0·002, CaCl₂ 0·012, MgCl₂ 0·06. Of these ions the presence of magnesium is not essential for the functioning of the elasmobranch heart.

The dogfish's heart reacts to ionic changes in a manner very similar to the amphibian heart, and so, indeed, do the other Fish hearts that have been investigated, e.g. eel, lamprey. It is, however, dangerous to make any sweeping generalisations because detailed analysis often reveals striking differences in the hearts of closely related species. For example, Mines[26] found that much smaller concentrations of $MgCl_2$ arrested the heart of *Raja* than that of *Scyllium*.

The possibilities of erratic variation in this respect are shown by the existence of one species of fresh-water Elasmobranch; little information is available concerning this animal, but its power of maintaining the saline composition of its blood must differ completely from the ordinary Elasmobranchs, for when these are put into fresh water they suffer a continuous loss of salts from the blood, and the osmotic pressure of the blood sinks until the animal dies in a few hours. Dakin[7] found that the osmotic pressure of the blood of a dogfish placed in fresh water sank from $\Delta = -1\cdot9°$ C. to $\Delta = -1\cdot453°$ C. in four hours.

THE ACTION OF IONS ON INVERTEBRATE HEARTS

Fredericq[14] divided the animal world as follows:

(1) *Coelenterates*

In these animals the fluids and the body tissues have the same qualitative and quantitative saline content as the surrounding water.

(2) *Marine Invertebrates*

In these animals the body fluids have the same qualitative composition as the sea water in which the animal lives. The body tissues, however, have a constant saline content, but their osmotic pressure is the same as that of the surrounding sea water and variations in the osmotic pressure of the sea water are compensated for by the cells varying their content of such organic constituents as urea.

(3) Elasmobranchs

The body fluids as well as the body tissues have a constant saline content, but the osmotic pressure varies with changes in the osmotic pressure of the surrounding sea water, and these variations are effected by changes in the content of organic constituents, urea being the most important constituent in this respect.

(4) Fresh-water and Land Invertebrates. Teleosteans and all Higher Vertebrates

In these animals the body fluids have a constant saline composition and both the quantitative and qualitative compositions are totally different from those of the water in which they live.

These general rules summarise the most striking facts about the composition of the body fluids in Invertebrates. They are not strictly true, however, for careful analyses show frequent exceptions.

For example, as regards the Coelenterates, Macallum (24) found that the saline composition of the *Medusa aurelia* was not identical with that of sea water since the Medusa contained 10 per cent. less magnesium and 30 per cent. less sulphates.

Again, as regards the Arthropods, Macallum (24) found that the blood of *Limulus* had a saline composition practically identical with that of sea water, but that on the other hand the blood of the lobster contained only one-seventh as much magnesium as the sea water in which the animal lived.

Since the blood of Invertebrates contains proteins it follows that, if it were separated from the surrounding sea water by a membrane freely permeable to ions but impenetrable to proteins, the total anion or kation content of the blood would not be identical with that of sea water owing to the Donnan equilibrium effect. This effect cannot, however, account for a difference in the ratio of the kations present in the two fluids. For example, the Donnan equilibrium effect might render the total kation content of the blood lower than that of sea water, but it could not cause a difference in the ratio of the concentrations of Na, K, Ca and Mg in the two fluids.

It will be seen at once that wide variations in the response of invertebrate hearts to ions are bound to occur, since in some animals the circulating fluid is of more or less constant composition, whilst in others its composition varies with variations in the surrounding water. The hearts of the latter may be expected to be the less sensitive to changes in the ionic composition of fluids surrounding them.

MOLLUSCAN HEARTS

A. *Fresh-water Molluscs*

Phillipson[27] found that the osmotic pressure of the blood of fresh-water Molluscs varied from $\Delta = -0.13°$ C. to $-0.23°$ C. Koch[21] found that in *Anodon* blood $\Delta = -0.09°$ C. He found that the heart beat well when perfused with a mixture of one volume of frog's Ringer to two or four volumes of water, and that the heart was arrested by the following molar concentrations of the various ions: NaCl 0.16, KCl 0.04, $CaCl_2$ 0.4, $MgCl_2$ 0.2. The heart required the presence of KCl and $CaCl_2$ in order to function, and excess of KCl arrested the heart in systole.

In most Molluscs excess of potassium causes systole of the heart; this is the opposite effect to that produced by excess potassium on the vertebrate heart but resembles the action of potassium on vertebrate smooth muscle. In chapter III it was pointed out that in most of their properties the molluscan heart muscles resembled vertebrate smooth muscle rather than vertebrate heart muscle.

B. *Land Molluscs*

Several observers have studied the heart of *Helix*. Evans[11] used frog's Ringer, and Cardot[3] and Hogben[20] used fluids very similar.

Evans[11] thought that the presence of potassium was not essential for the normal functioning of the heart, but both Cardot[3] and Hogben[20] found that absence of potassium arrested the heart in diastole. Excess of potassium arrests the heart in systole. Lack of calcium arrests the heart in systole, while excess of calcium arrests the heart in diastole (Hogben[20]). The heart is arrested in systole by acid at a P_H below 5.2.

On the whole, the hearts of fresh-water and land Molluscs react to ionic changes in a manner very similar to vertebrate plain muscle.

C. Marine Molluscs

The osmotic pressure of the blood of marine Molluscs is in all cases the same as that of the surrounding sea water, and the chemical composition of the blood appears to be very similar to that of sea water, since in most cases the hearts will function well when perfused with sea water.

The following perfusion fluids have been used by various workers:

Table 49

Molecular concentrations of perfusion fluids

	Pecten		Aplysia	Octopus	
	(Mines[26])	(Hogben[20])	(Heymans[17])	(Fredericq[14])	(Fry[15])
NaCl	0·42	0·4	0·52	0·59	0·43
Ca salts	0·0042	0·008	0·0056	0·0018	0·13 (CaSo$_4$)
Mg salts	0·083	0·08	0·064	—	0·03
K salts	—	—	0·0114	0·0013	0·044
NaHCO$_3$	—	—	0·0024	0·0012	—
C_H	10$^{-?}$	—	—	—	—

In *Pecten* heart Mines[26] found that Na, Ca and Mg were essential, but that K was not essential. He also found that the heart beat best with a neutral perfusion fluid, and that a feeble alkalinity (P_H 8·3) arrested the heart.

Hogben[20] found that lack of calcium arrested the heart of *Pecten* in diastole and that excess arrested it in systole.

Heymans[17] found that Na, Ca and K were all essential for the *Aplysia* heart and that excess of calcium or lack of potassium produced slowing and arrest in diastole, whereas lack of calcium or excess of potassium produced acceleration and arrest in systole.

In the cephalopod heart Fredericq[14] found that Ca and K were essential, but that Mg was not essential. Lack of calcium produced acceleration followed by arrest in a few minutes. Lack of potassium produced arrest in a short time.

The information about molluscan hearts is scanty, but it suggests that most of these hearts react to ionic changes in a manner more resembling the reaction of vertebrate plain muscle than the reaction of the vertebrate heart.

There is a striking difference between non-marine and marine forms, in that the presence of magnesium is essential for the activity of the hearts of the latter. In the case of *Pecten* the presence of potassium is not essential and the heart can function with a mixture of sodium, calcium and magnesium. These ionic requirements are very different from anything seen in vertebrate tissues.

The hearts of Tunicates appear to be similar to those of Molluscs in their ionic requirements. Tunicate blood has the same quantitative and qualitative content of inorganic salts as sea water, and the tunicate heart functions when suspended in sea water.

ARTHROPOD HEARTS

Marine Forms

In the marine Arthropods the osmotic pressure of the blood is the same as sea water, and the ionic composition of the blood appears to be similar, but is not identical in all cases.

Limulus

The heart of *Limulus* functions well when perfused with sea water. Macallum has shown that the inorganic constituents of *Limulus* blood are identical quantitatively with those of sea water.

Carlson (4) studied the action of ions both on the heart ganglion and on the heart muscle of *Limulus*. He found that potassium excess first excited and then paralysed the heart ganglion, whereas calcium excess paralysed the heart ganglion without initial excitation. Lack of Ca and K caused the activity of the ganglion to cease in about 1 hour. Calcium and potassium both produced a purely depressor action on the heart muscle. Lack of both calcium and potassium (i.e. pure NaCl) excited the heart muscle, and caused the muscle isolated from the ganglion to contract automatically. Acids stimulated the ganglion and depressed the muscle. Alkalies stimulated both ganglion and muscle.

Meek (25) measured the time required to produce paralysis in the various structures in *Limulus* heart when excess of potassium or calcium was added, and obtained the following results:

	Automatism of ganglion paralysed	Excitability and conductivity of nerve plexus paralysed	Excitability and conductivity of muscle paralysed
CaCl$_2$ 0·3 normal	90 mins.	150 mins.	211 mins.
KCl 0·3 normal	30 secs.	12 mins.	180 mins.

These figures show that the ganglion is far more susceptible to excess of potassium than the other tissues in the heart and that the muscle is much less sensitive than the nerves to ionic changes.

CRUSTACEAN HEARTS

The blood of Crustacea is isotonic with sea water, but Macallum has shown that the composition of the two is not identical, since the blood of the lobster only contains about one-sixth as much magnesium as does sea water.

Hogben[20] found that the following solutions (molar conc.) were suitable for the perfusions of the hearts of *Maja* and *Homarus*.

	Maja	*Homarus*
NaCl	0·5	0·5
$CaCl_2$	0·005–0·012	0·04
KCl	—	0·005–0·012

He found that substitution of one-quarter of the NaCl by urea somewhat improved the beat.

The ionic requirements of *Maja* heart are therefore extremely simple, since it can function in a mixture containing sodium and calcium alone.

It is interesting to note that the ionic requirements of *Homarus* and *Maja* are not identical. Moreover, Hogben found a considerable individual variation in the amount of calcium needed by the *Maja* hearts.

In both these species excess of potassium caused systolic arrest, and removal of calcium produced the same effect in *Maja*.

In *Homarus* the effect produced by lack of calcium depended on the concentration of potassium present. If the potassium content was 0·005 molar or less, then lack of calcium produced systolic arrest, but if the potassium content was 0·012 molar, then lack of calcium caused diastolic arrest.

These hearts were extraordinarily insensitive to changes in the reaction of the fluid, and would function well in a fluid with P_H 5·6; indeed, the *Maja* heart still functioned in a fluid with P_H 4·0.

This is in very striking contrast to the behaviour of vertebrate heart muscle, which is extremely sensitive to changes in reaction.

Fresh-water Crustacea

The osmotic pressure of *Astacus* blood is $\Delta = - 0.8°$ C. (Fredericq [13]). An analysis by Dohrn (1866) gives the following molar composition: NaCl 0.14, KCl 0.038, CaCl$_2$ 0.043, MgCl$_2$ 0.01. The figures for potassium and calcium are surprisingly large. Griffiths [16] gives the following molar ratios: Na 100, K 6, Ca 4, Mg 1, and these appear more probable.

Hoffmann [19] found that the heart beat well in a double concentration of frog's Ringer; this fluid has the following molar concentration: NaCl 0.22, KCl 0.004, CaCl$_2$ 0.002, NaHCO$_2$ 0.002.

He found that excess of either potassium or calcium produced little effect on the heart.

TRACHEATA

The osmotic pressure of insect blood varies from $\Delta = - 0.5$ to $- 1.0°$ C.

In water beetles Backman [1] found figures ranging from $\Delta = - 0.49$ to $- 0.95°$ C. He showed that if Insects were kept in dry air or in concentrated brine the osmotic pressure might rise to $\Delta = - 1.8°$ C. Polimanti [28] found that the body fluids of *Bombyx mori* (larvae and moths) had an osmotic pressure corresponding to $\Delta = - 0.75°$ C.

Griffiths has furnished analyses of the blood of a number of Lepidoptera. He found a salt content of about 3.6 per cent. This indicates an osmotic pressure of about $- 2.3°$ C. which seems improbable, since all other observations on insect blood indicate a saline content of about 1.0 per cent. ($\Delta = - 0.6°$ C.).

Griffiths found the following molar ratios: Na 100, K 6, Ca 4, Mg 2.

Dr Seliškar examined the reaction to ionic changes of the hearts of the stick insect (*Carausius*), and of the larva of *Cossus cossus*. He has kindly permitted me to mention his results which are not yet published. These hearts beat well when irrigated with a fluid of the following molar composition: NaCl 0.22, KCl 0.002, CaCl$_2$ 0.001 ($\Delta = - 0.8°$ C.). Removal of calcium arrested the hearts in a few minutes, but removal of KCl produced no visible effect in 2 hours. Increase of KCl to 0.02 molar produced slowing of the rate of conduction in the heart and arrested it in a few minutes. Increase of $CaCl_2$ to 0·01 molar at first increased the frequency and then produced arrest. Reduction of NaCl to 0·11 molar reduced the frequency of the heart.

THE REACTION OF ARTHROPOD HEARTS TO IONIC CHANGES

The data available are too scanty to permit of many generalisations, but they indicate certain contrasts between arthropod and vertebrate heart muscles.

The presence of potassium seems much less important for arthropod than for vertebrate hearts, and in many cases the heart functions normally in its absence, but in all cases the presence of calcium is essential. Excess of potassium produces arrest. Finally, the arthropod hearts appear to be very insensitive to changes in reaction. In general these characteristics resemble those of vertebrate skeletal muscle more nearly than those of vertebrate, plain or cardiac muscle.

THE ELECTROLYTE CONTENT OF THE BLOOD

The relative molar concentrations of the blood of a number of animals are shown in Table 50.

These figures show that although the total salt content of the blood varies from a quantity equal to that of sea water in the marine Invertebrates, to about one-fifth of this amount in Amphibia, yet the relative concentrations of sodium, potassium and calcium remain extremely constant. Most bloods contain a somewhat higher proportion of potassium than is present in sea water.

As regards magnesium, this is present in the same proportion as in sea water in most marine Invertebrates, but in some of these and in all other forms of animal life the proportion of magnesium in the blood is less than one-sixth of that in sea water.

A very interesting explanation of these differences has been supplied by Bunge and also Macallum who suggested that the ionic concentration of the blood of land Vertebrates was the same as that of the sea from which these animals originally emerged.

Table 50

Osmotic Pressure and Saline Content of Blood

	Δ of blood or serum	Ratio of molecular concentrations of ions in serum				Author
		Na	K	Ca	Mg	
Sea water	− 2·3 (18)	100	2·1	2·28	11·5	Macallum (24)
Marine Invertebrates	− 2·3 (18)	—	—	—	—	—
Limulus	—	100	3·3	2·33	10·6	Macallum
Homarus (Lobster)	—	100	2·2	2·72	1·62	,,
Elasmobranchs	− 2·26 (36)	—	—	—	—	—
Scyllium (dogfish)	—	100	2·7	1·55	2·32	Macallum
Marine Teleosts	− 0·761 (36)	—	—	—	—	—
Gadus (cod)	—	100	5·6	2·28	1·33	Macallum
Marine Mammals	− 0·763 (36)	—	—	—	—	—
Land Mammals	− 0·601 (36)	100	3·45	1·5	0·76	(38)
Birds	− 0·63 (36)	100	14·8	2·0	1·5	(38)
Land and fresh-water Tracheata	− 0·5–1·0 (37)	100	6·3	3·7	1·8	Griffiths (16)
Ganoid Fishes	− 0·64 (36)	—	—	—	—	—
Fresh-water Teleosts	− 0·545 (36)	—	—	—	—	—
Marine Reptiles	− 0·637 (36)	—	—	—	—	—
Land Reptiles and Amphibia	− 0·477 (36)	100	4·7	2·3	1·65	Urano (34)
Land and fresh-water Molluscs	− 0·1–0·4 (18)	—	—	—	—	—

The circulation of water from the sea to land as rain and back through the soil to the ocean is a process of continuous extraction of the soluble salts of the land, and hence the saline content of the ocean must have increased steadily since its original formation.

The potassium brought to the sea tends to be fixed by vegetables and animals, and the plant life on land fixes a considerable proportion of the potassium in the soil. Calcium is also continuously being removed by marine animals to form their shells. Hence the ratio of potassium and calcium to sodium was probably higher in previous geological periods than it is to-day.

These assumptions will explain why the blood plasma of Vertebrates has a much lower saline content than sea water and why the salts of their blood serum contain relatively more potassium and sodium. The most outstanding fact about vertebrate blood, however, is its low content of magnesium, and therefore the hypothesis outlined above necessitates the assumption that in the Cambrian age when the vertebrate forms are believed first to have left the sea the magnesium content was far lower than it is to-day.

Macallum(24 a) explains his figures by suggesting that marine Invertebrates which have always inhabited the sea have completely adjusted their body fluids to their external *milieu*, and hence the blood of *Limulus* has the same qualitative and quantitative composition as sea water. The lobster he suggests is a descendant of animals which had a lacustrine existence for some geological periods and hence the saline content of their blood still differs from that of sea water.

The Elasmobranchs he supposes to have always inhabited the sea but to have stabilised the quantitative and qualitative saline composition of their blood at an early period but later than the Cambrian period. He suggests that the Teleosts left the sea about the Cambrian period, and returned from a lacustrine to a marine existence, at some later period when the saline content of their serum was completely stabilised. The Mammals are supposed to have left the sea at the Cambrian period, and presumably the Amphibia and Reptiles must either have emigrated to land at an even earlier period or have stabilised the ionic content of their blood during a lacustrine existence. The high potassium content of Birds' blood seems, however, very difficult of explanation on this theory.

The saline content of Tracheata would agree with their having left the sea at a somewhat later period than Mammals.

In view of the marked power of Molluscs and Worms to adapt themselves to changes in the saline composition of their environment, it seems unsafe to speculate on the date of their origin as land and fresh-water forms.

The hypothesis put forward above is an attractive and interesting one, but far more detailed knowledge is needed regarding the saline content of the blood of Invertebrates before it will be safe to accept any wide generalisations regarding the relation between the saline content of the internal and external *milieu* of animals.

References

(1) Backman. *Pflüger's Arch.* **149**, 93. 1912.
(2) Beresin. *Pflüger's Arch.* **150**, 549. 1913.
(3) Cardot. *Compt. Rend. Soc. de Biol.* 1921, 2, 813 and 1922, 2, 1113.
(4) Carlson. *Ergebn. d. Physiol.* **8**, 441. 1907.

142 INFLUENCE OF ENVIRONMENT ON HEART'S ACTIVITY

(5) Clark. *Journ. of Physiol.* 47, 66. 1913.
(6) Cousy. *Arch. intern. de Physiol.* 21, 90. 1923.
(7) Dakin. *Biochem. Journ.* 3, 473. 1908.
(8) Dale and Thacker. *Journ. of Physiol.* 47, 493. 1913.
(9) Daly and Clark. *Journ. of Physiol.* 54, 367. 1921.
(10) Dennis. *Amer. Journ. of Physiol.* 17, 35. 1908.
(11) Evans. *Zeit. f. Biol.* 59, 397. 1912.
(12) Franca. *Arch. di Fisiol.* 7, 69. 1909.
(13) Fredericq. *Bull. Acad. de Belg.* (3) 25, 831. 1898.
(14) Fredericq. *Arch. intern. de Physiol.* 14, 126. 1913.
(15) Fry. *Journ. of Physiol.* 39, 184. 1907.
(16) Griffiths. *Physiology of Invertebrates.* 1892.
(17) Heymans. *Arch. intern. de Pharm.* 28, 337. 1923.
(18) Höber. *Physikalische Chemie d. Zelle u. Gewebe,* 1, 34 ff. Leipzig, 1922.
(19) Hoffmann. *Zeit. f. Biol.* 59, 297. 1912.
(20) Hogben. *Quart. Journ. Exp. Physiol.* 15, 263. 1925.
(21) Koch. *Pflüger's Arch.* 166, 281. 1917.
(22) Kolm and Pick. *Pflüger's Arch.* 185, 235. 1920.
(23) Lussana. *Arch. intern. de Physiol.* 11, 1. 1911.
(24) Macallum. *Proc. Roy. Soc.* B, 82, 602. 1910.
(24 a) Macallum. *Physiol. Rev.* 6, 316. 1926.
(25) Meek. *Amer. Journ. of Physiol.* 21, 230. 1908.
(26) Mines. *Journ. of Physiol.* 43, 467. 1911.
(27) Phillipson. *Arch. intern. de Physiol.* 9, 460. 1910.
(28) Polimanti. *Bioch. Zeit.* 70, 74. 1915.
(29) Pond. *Journ. of Gen. Physiol.* 3, 807. 1921.
(30) Sakai. *Zeit. f. Biol.* 64, 1. 1914.
(31) Sakai. *Zeit. f. Biol.* 64, 505. 1914.
(32) Schultz. *Amer. Journ. of Physiol.* 22, 133. 1908.
(33) Seliškar. *Journ. of Physiol.* 61, 172. 1926.
(34) Urano. *Zeit. f. Biol.* 50, 212. 1908, and 51, 482. 1908.
(35) Waller. *Journ. of Physiol.* 48; *Proc.* xlvi. 1914.
(36) Winterstein. *Hand. d. vergl. Physiol.* 1, 1288. 1925.
(37) Winterstein. *Hand. d. vergl. Physiol.* 1, 762. 1925.
(38) Winterstein. *Hand. d. vergl. Physiol.* 1, 118. 1925.
(39) Zwaardemaker. *Pflüger's Arch.* 173, 28. 1918.

Appendix I

THE PULSE FREQUENCY AND METABOLIC RATE OF VARIOUS MAMMALS AND BIRDS

(The metabolic rates recorded were in most cases taken at
room temperature—12–18° C.)

MAMMALS

Species	Body weight in gm.	Pulse frequency per minute	Metabolic rate in calories per kilo body weight per 24 hours
Vesperugo pipistrellus	4·3	660 (10)	—
Plecotus auritus	9·4	750 (10)	—
Muscardinus avellanarius (dormouse)	14	610 (8)	—
,,	17	590–670 (26)	—
,,	23	580–780 (26)	—
Mus musculus (common mouse; tame)	18	730 (7)	—
,,	25	600 (11)	400 (18)
,,	34	670 (7)	—
Young mice	1·75	—	977 (29)
			780 (20)
,,	3·8	—	660 (20)
,,	12	670 (7)	639 (32)
Mus norwegicus (common rat; tame)	200	360–520 (26)	227 (32)
		420 (11)	
Young rat	95	—	255 (12)
Sciurus vulgaris (squirrel)	222	320 (33)	—
Cavia porcellus (guinea-pig)	300	290 (22, 30)	—
		300 (9)	
,,	400	267 (3, 19)	94 (3)
,,	600	—	153 (32)
Young guinea-pig	50	—	286 (29)
,,	66	—	234 (20)
,,	145	—	168 (20)
Erinaceus europaeus (hedgehog)	520	300 (10)	—
,,	911	189 (33)	—

MAMMALS, contd

Species	Body weight in grm.	Pulse frequency per minute	Metabolic rate in calories per kilo body weight per 24 hours
Lepus cuniculus (tame rabbit)	1,434	220 (33)	—
,,	2,000	205 (17)	52 (32) 61 (28) 72 (23) 75 (33) 78 (3)
Young rabbits	40	—	355 (20)
,,	75	—	274 (20)
Lepus europaeus (hare)	2,500	64 (25)	—
Felis domesticus (domestic cat)	1,312	240 (33)	—
,,	2,500	116–128 (27) 110–130 (21)	79·5 (23)
Kittens	117	300 (11)	184 (20)
,,	201	—	170 (20)
,,	300	—	137 (20)
Arctomys marmotta (marmot)	3,600	186 (16)	—
Canis familiaris (dog)	3,200	—	91 (28) 88 (32) —
,,	5,000	105–125 (27)	— — 51 (24)
,,	6,500	120 (9)	68 (28) 66·1 (32) —
,,	9,600	96 (33)	61 (28) 65 (32) 40 (24)
,,	15,000	72–82 (27)	— — 35 (24)
,,	20,000	85 (9)	48 (28) 46 (32) 32 (24)
,,	30,000	—	38 (28) 36 (32) —
Puppies	400	—	145 (20)
,,	500	—	100 (20)
,,	650	—	111 (20)
,,	1,750	180 (11)	—
Capra hircus (goat)	33,000	135 (15) 70–90 (27) 80–100 (11)	49·5 (1)
Ovis aries (sheep)	50,000	70–80 (31)	27·4 (20)
Man	50,000	—	29 (2)
,,	66,000	55 (4)	25 (4) 32 (34) 38·7 (28)
,,	75,000	—	24 (32) 33 (23)
,,	80,000	—	22 (2)
Infants	1,900	129 (6)	95 (6)
,,	5,000	119 (6)	71 (6)
,,	10,000	106 (6)	51 (6)

MAMMALS, *contd*

Species	Body weight in grm.	Pulse frequency per minute	Metabolic rate in calories per kilo body weight per 24 hours
Sus domesticus (pig)	130,000	70–86 [27]	19 [34]
Equus caballus (horse)	380,000	55 [33]	—
,,	450,000	36–40 [31]	11·3 [34]
		39 [13]	15 [29]
			20 [20]
			27 [23]
Bos taurus (ox)	500,000	43 [5]	24 [5]
Calf	115,000	—	37 [20]
Elephas indicus (elephant)	3,000,000	46–50 [14]	—

References

(1) Barcroft et al. *Quart. Journ. of Med.* 13, 35. 1920.
(2) Benedict. *Journ. of Biol. Chem.* 20, 263. 1925.
(3) Benedict. *Journ. of Biol. Chem.* 20, 301. 1925.
(4) Benedict et al. *Publ. of Carnegie Inst. of Wash.* 280. 1919.
(5) Benedict and Ritzman. *Publ. of Carnegie Inst. of Wash.* 324. 1923.
(6) Benedict and Talbot. *Publ. of Carnegie Inst. of Wash.* 201. 1914.
(7) Buchanan. *Journ. of Physiol.* 37; *Proc.* lxix. 1908.
(8) Buchanan. *Journ. of Physiol.* 40; *Proc.* xliii. 1910.
(9) Buchanan. *Science Progress*, 5, 60. 1910.
(10) Buchanan. *Journ. of Physiol.* 42; *Proc.* xxi. 1911.
(11) Clark. Unpublished observations.
(12) Donaldson. *The Rat.* Philadelphia, 1924.
(13) Ellinger. Quoted from Tigerstedt. *Physiol. d. Kreisl.* 2, 468. 1922.
(14) Evans. *Elephants and their Diseases.* Rangoon, 1910.
(15) Gley and Quinquaud. *Arch. Néerl. de Physiol.* 7, 392. 1922.
(16) Hecht. *Zeit. f. d. ges. exp. Med.* 4, 259. 1915.
(17) Hering. *Pflüger's Arch.* 60, 425. 1895.
(18) Hill and Hill. *Journ. of Physiol.* 46, 81. 1913.
(19) Keilson. Quoted from Stübel. *Pflüger's Arch.* 135, 249. 1910.
(20) Kestner and Plant. Winterstein's *Hand. d. verg. Physiol.* 2, ii. 1037. 1924.
(21) Kirk. *Diseases of the Cat.* London, 1925.
(22) Koenigsfeld and Oppenheimer. *Zeit. f. d. ges. exp. Med.* 28, 106. 1922.
(23) Loewy. Quoted from Krogh. *The Respiratory Exchange of Animals.* London, 1916.
(24) Lusk and Du Bois. *Journ. of Physiol.* 56, 213. 1924.
(25) MacWilliam. *Proc. Roy. Soc.* B, 53, 464. 1893.
(26) Oppenheimer. *Zeit. f. d. ges. exp. Med.* 28, 96. 1922.

(27) Reichert. Inaug. Diss., Giessen. Quoted from *Zentralbl. f. Biochem. und Biophys.* **10**, 170. 1910.
(28) Rubner. *Zeit. f. Biol.* **19**, 535. 1883.
(29) Rubner. *Biochem. Zeit.* **148**, 222. 1924.
(30) Schott. *Arch. f. exp. Path. u. Pharm.* **87**, 309. 1920.
(31) Sisson. *Anatomy of Domestic Animals.* Philadelphia, 1914.
(32) Tigerstedt. Quoted from Krogh. *The Respiratory Exchange of Animals.* London, 1916.
(33) Vierordt, K. *Grundriss d. Physiol. d. Menschen,* 5th ed. p. 162. 1877.
(34) Voit, E. Quoted from Krogh. *The Respiratory Exchange of Animals.* London, 1916.

BIRDS

Species	Body weight in grm.	Pulse frequency per minute	Metabolic rate in calories per kilo body weight per 24 hours
Parus major (great titmouse)	14	870 (10)	—
Carduelis elegans (goldfinch)	16	920 (2)	920 (2)
Serinus canarius (canary)	20	1000 (2)	1060 (6) 860–880 (2)
Erithacus rubecula (robin)	20	—	1150 (6)
Ruticilla phoenicurus (redstart)	20	890 (10)	—
Fringilla coelebs (chaffinch)	15	700 (10)	—
,,	22	—	1000 (6)
Pyrrhula europoea (bullfinch)	23	—	940 (2)
Fringilla montifringilla (bramblefinch)	22	900–920 (10)	—
Passer domesticus (sparrow)	20	640–910 (10)	755 (13)
,,	24	800 (2)	890 (2)
Ligurinus chloris (greenfinch)	26	740 (2)	855 (2)
,,	—	700 (10)	710 (6)
Alcedo ispida (king-fisher, young)	42	440 (2)	—
Turdus merula (blackbird)	58	390–590 (10)	—
Sturnus vulgaris (starling)	74	—	270 (6)
Tinnunculus alaudarius (kestrel)	159	367 (12)	—

BIRDS, *contd*

Species	Body weight in grm.	Pulse frequency per minute	Metabolic rate in calories per kilo body weight per 24 hours
Corvus monedula (jackdaw)	140	342 (12)	—
Tame pigeon	237	244 (12) 141 (1)	—
,,	300	185 (2) 200 (1)	250 (2)
Corvus frugilegus (rook)	341	380 (3)	—
Corvus cornix (crow)	360	378 (12)	127 (6)
Syrnium aluco (wood owl)	360	—	106 (6)
Larus marinus (black-backed gull)	388	401 (12)	—
Psittacus erithacus (parrot)	430	320 (2)	—
Buteo vulgaris (buzzard)	658	301 (12) 282 (14)	93 (5)
Falco peregrinus (falcon)	960	347 (12)	—
Milvus ictinus (kite)	950	258 (12)	—
Domestic hen	1,000	354 (14)	110 (2) 93 (13)
,,	2,000	360 (1) 328 (8) 312 (12)	71 (15) 69 (11) 67 (13) 55 (5)
,,	3,000	304 (1)	65 (1)
Domestic duck	785	317 (1)	—
,,	1,000	—	90 (6)
,,	2,000	240 (2) 212 (12)	240 (2)
Anas boscas (wild duck)	770	195 (3)	186 (3)
,,	1,000	185 (3)	—
Domestic goose	2,800	144 (14)	—
,,	4,000	80 (9)	75 (7) 67 (15) 57 (11)
Gyps (vulture)	8,300	171 (12)	—
Domestic turkey	8,700	193 (12)	—
Arthropoides paradisea (Stanley crane)	11,000	120 (4)	—
Casuarius galeatus (cassowary)	60,000	70 (4)	
Struthio camelus (ostrich)	80,000	60–70 (4)	—

148 APPENDIX I

References

(1) Buchanan. *Journ. of Physiol.* **38**; *Proc.* lxii. 1909.
(2) Buchanan. *Science Progress,* **5**, 60. 1910.
(3) Buchanan. *Journ. of Physiol.* **47**; *Proc.* iv. 1913.
(4) Clark. Unpublished observations.
(5) Gerhartz. Quoted from Winterstein's *Handb. d. vergl. Physiol.* **2**, ii. 1037. 1924.
(6) Groebels. Quoted from Winterstein's *Handb. d. vergl. Physiol.* **2**, ii. 1037. 1924.
(7) Hari. Quoted from Winterstein's *Handb. d. vergl. Physiol.* **2**, ii. 1037. 1924.
(8) Keilson. Quoted from Stübel. *Pflüger's Arch.* **135**, 249. 1910.
(9) Mosso. *Arch. Ital. de Biol.* **35**, 21. 1901.
(10) Oppenheimer. *Zeit. f. d. ges. exp. Med.* **28**, 96. 1922.
(11) Rubner. *Zeit. f. Biol.* **19**, 535. 1883.
(12) Stübel. *Pflüger's Arch.* **135**, 249. 1910.
(13) Tigerstedt. Quoted from Krogh. *The Respiratory Exchange in Animals.* London, 1916.
(14) Vierordt, K. *Grundriss d. Physiol. d. Menschen,* 5th ed. 1877, p. 162.
(15) Voit, E. Quoted from Krogh. *The Respiratory Exchange in Animals.* London, 1916.

Appendix II

HEART RATIOS AND AORTIC CROSS-SECTIONS OF VARIOUS ANIMALS

These measurements were kindly supplied by Dr Beattie from the post-mortem room of the London Zoological Gardens, with the exception of those marked * which were made by the author.

MAMMALIA

Order and Species	Body weight in grm.	Heart weight in grm.	Heart ratio $\frac{\text{H.W.} \times 100}{\text{B.W.}}$	Aortic cross-section in sq. cms.
Quadrumana				
*Anthropopithecus troglodytes (chimpanzee)	22,600	128	0·57	1·23
Macacus rhesus	5,400	28	0·52	0·38
Macacus fascicularis	1,600	9·6	0·6	—
Callithrix jacchus	113	1·04	0·92	0·05
Carnivora				
*Felis domesticus (domestic cat)	1,750	7·5	0·43	0·13
,,	2,800	12·2	0·435	0·159
*New-born kittens (average of four)	126	0·76	0·6	0·017
*Canis familiaris (dog)	8,500	75	0·88	0·95
,,	12,000	102	0·84	1·14
,,	—	101	0·72	1·14
,,	21,000	155	0·74	1·2
*Puppy	4,500	41	0·9	0·63
Putorius putorius (polecat)	530	6	1·1	—
Procyon lotor	3,700	35	0·94	—
Mephites mephitica	2,400	14	0·59	—
*Phoca vitulina (common seal)	17,000	170	1·0	—
*New-born seal	5,450	88	1·63	0·79
Rodentia				
Castor canadiensis (beaver)	3,200	20	0·62	—

MAMMALIA, contd

Order and Species	Body weight in grm.	Heart weight in grm.	Heart ratio H.W. × 100 / B.W.	Aortic cross-section in sq. cms.
Rodentia, contd				
*Mus decumans (tame rat)	150	0·61	0·41	0·011
,,	180	0·74	0·41	0·018
,,	241	0·93	0·385	0·031
,,	500	1·6	0·32	0·033
*Mus musculus (tame mouse)	16	0·10	0·63	0·0026
,,	25	0·13	0·52	0·0038
,,	28	0·18	0·64	0·0050
Dasyprocta prymno-lopha	2,320	6·2	0·31	—
*Cavia porcellus (guinea-pig)	200	0·62	0·31	0·026
*Lepus cuniculus (tame rabbit)	2,000	5·8	0·29	0·096
Ungulata				
*Equus caballus (horse)	(610,000)	3,900	(0·64)	23·7
*Bos taurus (ox)	(590,000)	2,600	(0·45)	12·6
Capra siberica	25,500	170	0·67	0·94
*Ovis aries (sheep)	(45,000)	210	(0·46)	1·77
,,	(42,000)	193	(0·46)	1·76
*Sus domesticus (pig)	(45,000)	205	(0·45)	1·77
,,	(36,000)	161	(0·45)	1·76
Marsupialia				
Aepyprymnus rufescens	1,820	15·5	0·85	—
Dasyurus viverrinus	425	4	0·95	—

AVES

Order and Species	Body weight in grm.	Heart weight in grm.	Heart ratio H.W. × 100 / B.W.
Passeres			
Pyrrhocorax alpinus	226	4·35	1·9
Picariae			
Bucorvus abyssinicus	2,150	28·5	1·3
Guira piririgua	100	1·4	1·4
Colius affinis	57	0·35	0·61
Centropus rufipennis	240	1·8	0·75

AVES, *contd*

Order and Species	Body weight in grm.	Heart weight in grm.	Heart ratio H.W. × 100 / B.W.
Psittaci			
Ara macao	730	9·9	1·36
Palaeornis torquata	114	1·2	1·06
Psephotus multicolor	57	0·7	1·23
Eclectus roratus	340	4·9	1·44
Platycercus eximius	85	1·1	1·3
Lorius erythrothorax	175	2·0	1·1
Striges			
Bubo bengalensis	1,140	9·8	0·85
Athene noctua	142	1·7	1·18
,,	114	1·4	1·22
Accipitres			
Serpentarius serpentarius	3,000	34	1·1
Melierax monogrammicus	200	1·8	0·9
Archibuteo sancti-johannis	800	6·5	0·81
Otogyps calvus	3,000	23	0·77
Herodiones			
Leptoptilus javanicus	4,600	36	0·78
Leucophryx candidissima	255	2·2	0·87
Odontoglossae			
Phoenicopterus roseus	2,270	33	1·45
Phoenicopterus chiliensis	2,150	32	1·5
Anseres			
Anser albipons	2,170	26	1·2
Columbae			
Spilopelia tigrina	85	1·4	1·66
Goura victoriae	1,600	22	1·4
Columba guinea	255	3·3	1·3
Gallinae			
Pavo cristatus	5,000	39	0·78
Gallus sonnerati	1,020	8	0·8
Acryllium vulturinum	900	7·5	0·83
,,	1,140	7·5	0·65
Chryolophus amherstae	460	6·1	1·32
Pternistes nudicollis	455	3·1	0·68
Francolinus ponticerianus	227	1·6	0·7
Impennes			
Aptenodytes pennanti	9,800	74	0·76
Struthiones			
†*Struthio camelus*	70,000	765	1·08

(† The aortic cross-section of *Struthio camelus* was 4·9 sq. cm.)

Appendix III

VARIOUS FIGURES FOR AORTIC CROSS-SECTION

(Estimated figures in brackets)

Species	Body weight in grm.	Heart weight in grm.	Heart ratio $\frac{\text{H.W.} \times 100}{\text{B.W.}}$	Aortic cross-section in sq. cms.	
Mus decumans (tame rat)[1]	36·5	(0·22)[2]	—	0·0061	
,,	129	(0·59)[2]	—	0·0150	
,,	296	(1·09)[2]	—	0·0285	
Cavia porcellus (guinea-pig)[1]	156	(0·62)	—	0·0145	
	726	(2·2)	—	0·0455	
Man: New-born	3,000[3]	19[4]	0·63[4]	0·59[5]	0·53[6]
1–2 years	10,000[3]	54	0·54	1·0	1·10
5–10 years	20,000[3]	110	0·55	1·67	1·80
Adult	50,000	250	0·50	—	—.
	65,000	325	0·50	3·3	3·86
	100,000	410	0·41	—	—
Elephant (Indian) "Small female"[7]	(2,000,000)	10,500	(0·52)	63	
Average of 3 animals[8]	(2,500,000)	10,900	(0·44)	—	
Single animal[9]	2,000,000	19,000	(0·95)	—	
Cetacea[10] *Delphinapterus leucas* (white whale) 4×0·87 metres[11]	(0·77 ton)	—	—	45	
Monodon monoceros (narwhal) 4·6×0·83 metres[12]	(1 ton)	4,900	(0·49)	—.	
"Whale" 4 metres long[13]	—	22,000	—	57	

Appendix III, *contd*

Species	Body weight in grm.	Heart weight in grm.	Heart ratio $\dfrac{\text{H.W.} \times 100}{\text{B.W.}}$	Aortic cross-section in sq. cms.
Balaena mysticetus (Greenland whale) "Sucker" 5·8 × 1·22 metres (14)	(5 tons)	29,000	(0·6)	180
Hyperoodon Rostratus (bottle-nosed whale) 7·2 × 1·55 metres(15)	(4·6 tons)	(50,000)	(1·1)	79
Balaenoptera musculus (common Rorqual) 18·4 × 3·7 metres(16)	(45 tons)	(200,000)	(0·44)	374
Balaenoptera Sibbaldi (sulphur-bottom whale) 24 × 4·4 metres(17)	(80–100 tons)	—	—	710
Balaena mysticetus (Greenland whale) 18 metres long(18)	(70–80 tons)	—	—	730

References

(1) Dreyer, Ray and Walker. *Journ. of Physiol.* **44**; *Proc.* xiv, 1912.
(2) Estimates based on figures given by Donaldson. *The Rat.* Philadelphia, 1924.
(3) Figures from standard tables of normal weight of male children.
(4) Wideröe, Sotus. *Die Massenverhältnisse des Herzens unter pathologischen Zuständen.* Christiania, Dybwad, 1911. Quoted from *Handb. Normalen u. Path. Phys.* **17**, 145. 1926.
(5) Suter. Quoted from Junk. *Tabulae Biologicae,* **1**, 116. 1925.
(6) Thoma. Quoted from Junk. *Tabulae Biologicae* **1**, 115. 1925.
(7) Retzer. *Anat. Record.* **6**, 75. 1912.
(8) Evans. *Elephants and their diseases.* Rangoon, 1910.
(9) Gilchrist. Quoted by Evans (8).
(10) The weights of the whales have been calculated from the estimates of their volume. The weights are given in tons (1 ton = 1000 kg.). The measurements given of whales are the length, and the diameter at broadest part. Cf. Scoresby, *Journal of a Voyage to the Northern Whale Fishery,* p. 155. Edinburgh, 1823.
(11) Barclay and Neill. *Memoirs of the Wernerian Nat. Hist. Soc.* **3**, 371. 1821.
(12) Scoresby. *An account of the Arctic Regions,* Edinburgh, 1820. Vol. 1, p. 495.
(13) White and Kerr. *Heart,* **6**, 205. 1916.
(14) Scoresby. *Journal of a Voyage to the Northern Whale Fishery,* p. 151.
(15) Bouvier. *Ann. des Sci. Nat.,* 7th Ser., Zool. B, **13**, 288. 1892. (Heart weight calculated from dimensions 57 × 47 cm.)
(16) Murie. *Proc. Zool. Soc.* **33**, 206. 1865. (Heart weight calculated from dimensions 105 × 80 cm.)
(17) Turner. *Proc. Roy. Soc. Edinburgh,* **26**, 197. 1872.
(18) Hunter. *Phil. Trans. Roy. Soc.* 1787.

INDEX

Printed in the United States
By Bookmasters